screencraft

costume design

Bonnie
first meeting B...
B...
V.R.

587 956

Deborah Nadoolman Landis

screencraft

costume design

Focal Press is an imprint of Elsevier.

The publisher offers special discounts on all bulk orders of this book.
For information, please contact:

Manager of Special Sales

Elsevier

200 Wheeler Road

Burlington, MA 01803

Tel: 781-313-4700

Fax: 781-313-4882

For Information on other Focal Press publications, visit our World Wide Web home
page at: http://www.focalpress.com

ISBN 0-24080-590-9

10 9 8 7 6 5 4 3 2 1

contents

introduction

Costume design in Hollywood today follows a path virtually unchanged since its early roots in the theater—from initial meetings with the director, who is the interpreter of the written text, to the appearance of the actor in costume on stage. These creative steps—simple and formally prescribed—comprise a chronological order of events that seldom varies. Costume designers may modify this discipline slightly from film to film, but our methodology is so orthodox, it reflects international standards and practices in the film industry today. The process of costume design in film is one distinct unit in the building blocks of producing a film. Film is a collaborative art. If each physical frame of film provides a rigid proscenium for action, then cinematographers, art directors, and costume designers fill those dimensions as concretely as in a conventional theater. To the left or right of the motion picture camera, just beyond the frame of every film, stands a costume designer watching the movements of the action, the placement of the atmosphere (extras) and of the principal actors.

Behind this book lay years of frustration with the poor quality of the preponderance of existing literature documenting motion picture costume design. For the better part of the last hundred years, our history has been relegated to a shabby place on the sidelines of film scholarship. There have been a few books published recently, their titles exploiting "Hollywood fashion," or the microscopically examined wardrobe of Hollywood stars.

At the root of the problem with existing film costume literature is a lack of understanding (or a basic misunderstanding) about what a costume is, and what it is not. Costumes are one tool a film director has to tell the story of the movie. Fashion and costume are not synonymous; they are antithetical. They have directly opposing and contradictory purposes. Costumes are never clothes. This is a problematic concept for fashion writers, designers, and magazine editors, and a real stumbling block to being able to understand costume design in film.

Lamentably, it is with the star that costume design literature begins. A vast majority of books on costume design mix up actors with the characters they play. This fusion of fact and fantasy, role and reality, seems to originate in the earliest writing on film stars. As a result of this intimate connection between the audience and actors, the role of costume designer disappears. Star power is so influential, and pervasive, it becomes impossible for the public to believe that an individual was responsible for designing the costumes. Storytelling, the original function of costume, is forgotten. This may explain why the role of costume design has been ignored by historians, and is deemed irrelevant and inconsequential by the fashion press.

The working process of the costume designer, therefore, lies undiscovered, rarely acknowledged by the film historian or by the film-going public. The most sophisticated audiences overlook some of the most subtle and effective contemporary costume design in film and television. As Cecil Beaton wrote in 1979: "It is true that Hollywood introduced a new sense of fashion and beauty to the world and, sad to say, also true that the designers of the clothes worn by the famous Hollywood stars were never really given their full credit." Bob Ringwood believes that if a costume designer gets too involved with fashion, he's not doing his job properly. Costumes are not designed to sell themselves. Stars are invented long before they show up on the now famous "red carpet," entering award shows wearing glamorous gowns. When designers create accurate characters for a film, those honest characterizations help propel actors into stardom. Sandy Powell clarifies that sentiment, "The most important thing about costume design, and the most exciting part, is helping create a character and contribute to a story. It's not about how glamorous, or sexy, or wonderful somebody looks. It's about making the character."

A discouraging reality for costume designers is the power of the fashion industry to co-opt our craft, by producing collection runway shows "inspired" by our designs. Many of the costume designers in this volume have influenced worldwide fashion trends, yet few have been given credit or authorship in the fashion press.

The opportunity to add to this limited field of study presented itself to me while I was in the last year of my doctoral dissertation in History of Design at the Royal College of Art in London. As President of the Costume Designers Guild, and as a 30-year veteran costume designer myself, I was in a position to contact some of the most brilliant film artists working today. As they shared their journeys with me, several commonalities emerged—similarities in perspective and philosophy that traverse age, gender, and culture. This constancy exists because creating truthful fictional characters through costume remains our mantra.

Costumes and characters for a script are defined and then painfully refined by the designer, up until the moment the camera has captured them irrevocably, and this process can be angst-ridden. Each frame of film is an empty canvas for the film designer. We collaborate with cinematographers and production designers to create credible environments for the narrative of each script. Nothing is left to chance, yet the scene, the people, and the place must appear to be genuine. Every choice of color, texture, and pattern for costumes and sets is a weighed decision by a partnership of designers.

If the production designer is responsible for the "where" of a film (Where are we? Where is the action taking place?), the costume designer is responsible for the "who" (Who is this character? What can the audience expect from this person?)

Before an actor speaks, his wardrobe has already spoken for him. In film jargon, this is considered "telegraphing" information. Costume designers are concerned passionately with one thing: creating characters that are truthful.

Designers suffer the effects of their own virtuosity when actors arrive on the set wearing costumes that may resemble their own street clothes. Gabriella Pescucci says that, "Regardless of the type of characters I'm dressing, my greatest satisfaction comes from having my work disappear in the film. If it doesn't jump out at the audience, it's serving the purpose of the story." A costume designer might be called the production designer of the body. We decorate the landscape of the written exposition, right below an actor's chin, where the dialogue is being delivered. Every garment worn in a theatrical production is a costume. In his book *Pretty Pictures* (1998) on film production design, C. S. Tashiro describes the role of costumes in almost architectural terms: "Costuming is the first circle of cinema's affective space. Yet it creates an immediate problem... (costumes) must be similar enough to the audience's appearance not to rise to the level of awareness, but also different enough to answer a narrative's specific needs. In the process, costume design creates an exchange value based on these contradictory, simultaneous experiences of emotional identification and objectivity."

Unlike fashion, which is designed for our three-dimensional world, costumes are designed to appear in two dimensions on film, which flattens and distorts the image. Motion picture costumes are designed to appear on one character, for one specific scene, in the emotional arc of a movie, lit and framed by the cameraman, reflecting an artistic choice. Costumes add essential information to the moment of a scene, of a story, to help achieve the visual and narrative goal of the filmmaker.

There is a basic, but devious, misapprehension about contemporary costume design, which many of the costume designers in this volume address with frustration. Like the popular myth about actors improvising their dialogue, contemporary costumes—everyday clothes—are taken for granted by the public and press. Contemporary costume design, whether in film or in television, is a cinema art form continually undervalued and misunderstood. The notion that contemporary costumes are "shopped" by designers, reaching the screen unaltered, with fashion designers' labels intact, is an oft-repeated urban legend with no substance. We are often asked the innocent question, "Where did you get it?" The answer is, we designed it. Designers clarify this confusion in many of their interviews. Albert Wolsky reflects that, "Designing contemporary films is difficult. My job is to identify, through elimination and simplification, who somebody is."

The basically unglamorous nature of costume design contributes to its obscuration in film literature. The process is a prosaic one: costume designers meet with the filmmaker, read the script, spend a great deal of time on research, what Anthony Powell calls, "Visually orchestrating a script, and finding the right musical key for the production." The designer and director share ideas, which may take the form of costume sketches, photographs, or actual costumes, then meet with the production designer, cinematographer, actors, and finally manufacture, rent, and/or purchase the costumes. The costume designer must first serve the story and the director, and it is the director who makes the ultimate decisions regarding the costumes. When a costume designer receives a script, the process of developing the visual shorthand for each character begins. Costume sketches, fashion research, and actual garments are used to help costume designers, directors, and actors develop a language to build each character.

Designers are engaged on a film from pre-production, through shooting until wrap, when the costumes are put away in storage, sold, or discarded. Designers can be separated from family and friends, sometimes for months, in isolated film locations around the world, and it can be lonely. Workdays run over 16 hours for a costume designer on any film. It is a grueling job, requiring vast reserves of patience and stamina. Costume design is the engine behind Hollywood glamour, and we work very hard to make it look so easy.

Although Jeffrey Kurland has designed many contemporary and period films, his approach to costume design is always the same: "Designing costumes is storytelling, in the same way that a writer or a director tells a story. As costume designers, we get under the character's skin the way an actor does." The more specific and articulate a costume is, the more effective it will be with an audience. The minute details of costume, often relished by actors, enhance their performances in imperceptible ways. Many actors credit their costume as a guide to the discovery of their character. Like everyone, actors often need sensitive designing for less than perfect bodies. Flattering figures, camouflaging flaws, and rebalancing proportions is often part of the designer's job description.

Comedy is another underrated venue for costume designers, where brilliant work is routinely accepted as a given. The reality of working on a comedy is that every costume is as carefully meditated upon as if it was a romantic lead in the most lavish Henry James adaptation. The difference is that a comedy—whether it is **Monty Python's The Meaning of Life** or **The Birdcage**—requires the designer to do much more than period research. The everyday business of a comedy film relies on the wit, humor, and clever resourcefulness of the costume designer. The pressure is on

the designer, who is frequently called upon by the director to enhance a moment or a scene. It is often up to the designer to come up with sight gags, fat suits, breakaway costumes, crazy hair, and makeup to develop wacky characters through subtle or outrageously funny costumes.

Costume is unique among film crafts because of its power over our cultural imagination. Costumes sometimes seem to magically "appear" or design themselves, or emerge fully formed from the collective global unconscious. I have found this to be especially true of costumes that instantly become iconographic, like my design for Indiana Jones' jacket and hat. When fueled by successful film storytelling, costume can transcend its original function, and take on a life of its own in the public embrace. Sometimes, unpredictably, a character can explode into film stardom for an actor. Individual costumes, whether Indiana Jones or Erin Brockovich, can independently, spontaneously, spawn worldwide trends. New "classics" feel like they have always been part of our culture, but behind every costume there is a costume designer.

Some of the costume designers included in this volume are close friends; there are others whom I look forward to meeting in person. I remain a fan, simply in awe of the beauty of their work. Anthony Powell and I had an interview that began at dawn and ended after dinner, but we easily could have talked all night and started over at breakfast again. Piero Tosi was a joy to be with for three hours. I tried to absorb as much as I could. Many of the designers were in production and on location when I contacted them for interviews. Everyone pitched in and responded with warm enthusiasm and kindness. My friend, film historian Alberto Farina, assisted and translated my interview with the great Piero Tosi, and interfaced with Gabriella Pescucci in Rome. I am grateful to him for his invaluable help here, and for his help on my doctoral dissertation interview of Vittorio Storaro as well.

The pool of incredible talent in the field of costume design is greater than it has ever been. The work of Theoni Aldredge, Colleen Atwood, Jenny Beaven, John Bloomfield, Chen Changmin, Judianna Makovsky, Maurizio Millenotti, Ruth Myers, Tom Rand, Penny Rose, Ritta Ryack, Huamiao Tong, and Julie Weiss could fill a companion volume, and I could name a dozen more to fill a third and fourth as well. These costume designers are creating and making history on film sets around the world. Sadly, last year we lost some of our very best, Danilo Donati and Shirley Russell; I regret missing those interviews. Their work will live on to inspire us all.

My thanks go to Erica ffrench at RotoVision in Brighton, who hired me to edit this volume on the response to an e-mailed inquiry about the future publication date of this book for my doctorate! Little did she realize how much I cherished the opportunity she graciously offered me. She has bravely and patiently waited while her experienced designer/neophyte writer assembled interviews from all over the world. Coordinating the visuals and sketches was Amanda Bown. She and Erica had the unenviable task of tracking down designers, who are, at best, moving targets. This left me free to be a labor leader, student, wife, and mother to Max and Rachel. Our assistant Sharon Dolin worked her usual magic acting as go-between. She has a genius for connecting people. Amanda Pisani, my over-qualified research assistant, was volunteered into working on interviews, while I was helping reinvent our union last year. There would be no book without her constant help and steady hand on the keyboard. She was always cheerleading, ever positive of my writing abilities, despite my whining about being "only a designer."

I would like to dedicate this book, and all my work, to my husband, John Landis. He is a filmmaker of the greatest humanity and humor, and a national treasure. Together, we have survived success, and have been lucky and unlucky. To quote his hero, Mark Twain, "Grief can take care of itself: but to get the full value of joy you must have somebody to divide it with." John's support and love are present and constant, every single day of my life.

Costume designers are passionate storytellers, historians, social commentators, humorists, psychologists, and magicians who can conjure glamour and codify icons. Costume designers are flexible but canny project managers who have to juggle ever-decreasing wardrobe budgets and film schedules, and battle the economic realities of worldwide film production. Ultimately, as Piero Tosi suggests, perhaps the essence of costume designing is the willingness and humility to accept each project as a new venture, to bring no preconceptions to the work, and to accept that each film is a learning process.

DEBORAH NADOOLMAN LANDIS, JUNE 2002

biography

James Acheson grew up in Essex in the UK and attended art school in Wimbledon before venturing into costume design. Early in his career, Acheson worked with members of the Monty Python team, designing witty costumes for Terry Gilliam's **Time Bandits** (1981), **Monty Python's The Meaning of Life** (1983), and creating the brutalist futuristic society for **Brazil** (1985). His collaboration with Bernardo

james acheson

interview

Bertolucci on **The Last Emperor** (1987) offered audiences an opportunity to experience a world that disappeared with the Chinese revolution, the mass numbers of costumes, which are unbelievable, and the breathtaking beauty of the court of the Forbidden City. In detail and quality, the costumes of **The Last Emperor** are simply unsurpassed. The film brought Acheson his first Academy Award for costume design and led to two other Bertolucci films: **The Sheltering Sky** (1990), and **Little Buddha** (1993). He won a second Oscar in 1988 for Stephen Frear's exquisite 18th-century satire **Dangerous Liaisons**. Acheson's third Oscar was earned for Michael Hoffman's wild 17th-century romp, **Restoration** (1996). He designed costumes for **Wuthering Heights** (Peter Kosminsky, 1992), **Frankenstein** (Kenneth Branagh, 1994), and **The Man in the Iron Mask** (Randall Wallace, 1998). In spite of these achievements, Acheson has managed to avoid being typecast as a period designer, and recently designed the costumes for Sam Raimi's **Spider-Man** (2002). Like all great costume designers, James Acheson has the ability to serve every kind of story.

My mother was the first person who influenced my career as a designer. She arranged for my school's art department to be available to me after hours so that I could study academic subjects, and yet not neglect my creative side. In terms of film design, I was greatly influenced by Piero Tosi's work and by the look of Italian films, such as those of Fellini and Visconti. As for Hollywood, I remember being fascinated by Busby Berkeley's movies when young. Later on, I became attuned to the designs of Banton, Adrian, Orry-Kelly, and Plunkett.

I grew up in Essex, and went to art school in London to study theater design. After my education was complete, a friend told me that the BBC costume department had several job openings. I got a job there as an assistant, and that is how I started as a costume designer. I was one of the few members of the staff who didn't object to working with fiberglass and plastic, so I was asked to design the costumes for a science-fiction show called *Doctor Who*. I designed 36 episodes. My budget was tiny, and I learnt all about being resourceful as a costume designer.

My first film job was for **Flash Gordon**, but the director was fired and both the art department and costume department were let go with him. Shortly thereafter I was introduced to Terry Gilliam. Gilliam asked me what I thought of his movie **Jabberwocky**, and I said, "I thought it was remarkable. I almost had to wash my hands after I'd seen it, it was so stinkingly realistic." We got on well, and he offered me the costume design job on **Time Bandits**. This was in 1979, and I was paid £94 a week.

If a script is any good, there is something very precious about reading it for the first time. I try to create a very special space in which to sit and read, and I try not to design the movie in my mind as I read. But the truth is, I'm always thinking about what the costumes should look like and how I would go about making them. I like to read the script at least three or four times—and in an ideal world, spend a day talking with the director.

Most of the films for which I've designed the costumes are period pictures; contemporary movies are difficult for me. I don't do them very often, and I don't think that I'm very good at them. Regardless of whether it's a period or contemporary picture, the costume design process is basically the same. When I designed the costumes for **The Last Emperor**, Bernardo Bertolucci said, "You cannot be unfaithful until you know the truth." What I think he meant was that if you throw yourself into the research and find the essence of the world in which you're designing, then you can really start to create; then you can start to break the rules and to be unfaithful, with confidence. You can start making choices, and you can bend those choices. Sadly, I don't think a costume designer often gets to work at that level; there isn't time to get that far.

My experiences on **Dangerous Liaisons** and **The Last Emperor** differed enormously. Bertolucci had seen my costume designs in **Brazil**, and he noticed a detail that nobody else has ever mentioned to me: I had built a striped suit for Ian Holm, with horizontal stripes on the jacket and vertical stripes on the pants. It was a subtle way to emphasise Holm's rather square physique. I think Bertolucci hired me for **The Last Emperor** on the basis of the striped suit. When we started working on the movie, they didn't have enough money to make it. We were all put on half-pay, and I had one assistant to work with in a tiny little room. But it was wonderful, because we had six months to really examine the period—China from 1903 to 1969. By the time we got to start building things, we had created an excellent database. However, because we had so little money, we would build a little bit and continue developing ideas. We learnt as we went along, and this gradual build-up of wardrobe meant that we weren't making mistakes very often. My crew and I had a real commitment to this movie. It went incredibly smoothly for a film that had such alarming logistics—over 10,000 costumes, shooting for 26 weeks in China.

It was truly a wonderful process, in terms of having a continuing relationship with the director. I had never before worked with a director who saw me as a collaborator. Bertolucci used to say to his crew, "You're all my collaborators. You are my lemons. Every day, I come and give you a little squeeze. Then I come back the next day, I give you another little squeeze. In fact, I squeeze every day until you are dry." And he would do exactly that.

We had about 50 extras in a marketplace in one scene, and there was one extra with a turban on, a little bit of a ragged

1

(1–4) **Brazil**: Directed by Terry Gilliam, the film is an Orwellian vision of the future. The plot looks forward from the 1940s to a sprawling, fascistic society ruled by an oppressive bureaucracy. The film's visual style is high fantasy. Jonathan Pryce is Sam Lowry, a worker in the Ministry of Information who gets caught by the system, and plots his escape in surreal dream sequences (3–4).

2

3

4

3

(1–7) **Dangerous Liaisons:** Acheson has mostly designed period films, and admits he finds designing for contemporary pieces more difficult. When creating the sumptuous costumes for Stephen Frear's 1988 adaptation of Choderlos de Laclos' 18th-century novel, Acheson had to travel to Europe to find suitable fabrics: "Everything I looked at in England was hideous. Nothing had that sort of buttery, sculptural quality." A tense, superbly acted film with powerful sexual overtones, Acheson calls working on location for it, "the happiest time I ever remember."

1

2

4

5

6

7

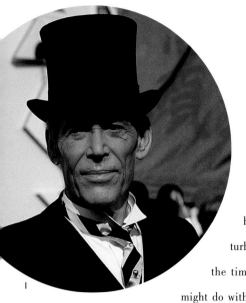

turban. Bertolucci said, "The man with the turban has such a beautiful head, can I take the turban off? Let's go look." He took the time to talk to me about what he might do with the 50th extra on the left. He delighted in details. He would say, "Taking the hem of a skirt up an inch can change the character."

One day I got a phone call to go to the set. I'd been there at five o' clock that morning, getting everyone dressed for shooting that day, and I was in the middle of preparing the 1,000 extras for the next day. But off I went, through Beijing at rush hour, to witness this huge scene with all these lamas with trumpets. Bernardo turned to me and said, "I want you to see your work." I told him that he had seen it that morning, and he replied, "No, you don't understand. This camera, my magic box, it records not only what I am shooting, but also the energies and the love, the passion, of all my creators. And I need you now and again, to be here. Not only to witness, but to somehow be part of this." Isn't that great? At the end of two years, he asked me, "What will you miss now you are going to work with another crew?" And I said, "I guess what I will miss most is your respect."

The director of **Dangerous Liaisons**, Stephen Frears, asked me to design the costumes on the basis of my work for **The Last Emperor**. We had only nine weeks to prepare. It was a situation where you just go for it. We set up a little workshop in London and divided the characters among us. I didn't know anything about the 18th century, so I dove into the research. We looked at lots of real clothes, but I was most impressed by the paintings of the era. Such effort had been taken with the

folds and fabric of the portraits, and it struck me that it was so important to find the right fabrics. Nothing I looked at in England was woven in the right way. Nothing had the sort of buttery, sculptural quality I needed. Most of the material we used came from France and Italy. We did find this beautiful bit of original 18th-century embroidered silk, and we had just enough to make the bodice of a dress for Glenn Close. We'd dyed the silk for the skirt and we got through two fittings, but the fabric couldn't withstand the handling and, by the time the dress arrived on the set, it was disintegrating. I knew that for continuity purposes, Glenn would have to wear it off and on for a month, and it was literally falling apart. It was a favorite of Glenn's, and I felt so badly telling her that we would have to reject it. One wonderful aspect of working on this movie was the opportunity to design for women of all ages, from 18 to 83, from Uma Thurman to Mildred Natwick.

Working on that film was a big adjustment. None of the actors knew a great deal about the 18th century, and they just let me take the lead. In a way, it felt great to be that powerful and to have so much control over my work. But at the same time, it was unsatisfying. The "working conditions" were also very different in that they were quite luxurious. We shot in beautiful châteaux, ate wonderful French food, and only worked on weekdays. The entire movie was shot in ten weeks.

Restoration, of course, gave me an opportunity to work with production designer Eugenio Zanetti and director Michael Hoffman. Zanetti is brilliant, and he had a unique sense of the world of the court of the restored Charles II. It was very elaborate, and very theatrical. Hoffman posed the most interesting question, asking all of us, "With all the trappings of this period, what kind of associations can you make with

3

2

4

(1–5) **The Last Emperor**: Acheson designed over 10,000 costumes for this startling epic on the life of Pu Yi, former Emperor of China. Working with the director Bernardo Bertolucci was a collaborative experience for Acheson. He was given a considerable amount of time to research the period (1903–1969), and as a result, the creative process suffered relatively few hitches. Pu Yi begins his reign as the Emperor on the Dragon Throne at the age of three (3), is forced to abdicate at age six and is tutored by his Scottish tutor (played by Peter O'Toole) (1,4) inside The Forbidden City. In exile, he becomes a playboy (2), and a pawn in the hands of the Japanese (5).

5

(1–3) **Spider-Man**: A largely faithful adaptation of the hugely popular Marvel comic, this film employed many technically brilliant special effects. The costumes of Spider-Man (1–2) and his evil enemy, the Green Goblin (3), had to be flexible, durable and incredibly photogenic—Acheson had eight people working on perfecting them.

1

2

3

the present world? How can you make this story accessible and not some stuffy costume drama?" I thought, "Great. We're in this period of *déshabillé*, where you wore things unbuttoned, in a sort of loose and louche way. We've got a king who's debauched and a hero who always seems to be putting his clothes on or taking them off. Let's loosen everything." Even with this sort of informal approach, the costumes were very showy. When you say to an actor, "Look, you're going to wear half a sheep on your head, have great big sleeves and four-inch heels, and I want you to give a performance as the King of England," you're posing a real challenge for him. It is very difficult to fight one's way through all the froth and furbelow of the mid-17th century. But Robert Downey Jnr. and Sam Neill were able to do it; they weren't dominated by the sets and the costumes. I think that this was one element of the movie—the interplay between the costumes and the actors' performances—that worked beautifully. The chance to work with these great actors was perhaps the best part about the experience for me. I would plunk great big, cheapo wigs on them, and get them to walk around and feel comfortable in them. Although the film looks incredibly lush, we had very little money for costumes.

For me, costume design is all about the development and support of a character. The essence of costuming is supporting the performer. One type of design that I really don't appreciate is what I call the "look at me" school—where the costumes are wearing the actors, rather than the actors wearing the costumes. That might sound strange coming from me, in light of the ostentatious work that I've done. Nonetheless, I think that costume designers' work is fundamentally to support the performer in his role, helping to create a world in which he exists.

I think that the craft of costume design has changed over the years. Filmmaking is becoming increasingly reliant on technology. The film stock and processes being used are more and more scientifically complex. For example, when I was designing the costumes for **Spider-Man**, I had to work with a green screen and think about accommodating harnesses and other equipment for the stunts. It limited my choices in terms of color and shapes. Today, when you work on Hollywood movies, you're very much a part of the film industry, with the emphasis on industry. Moviemaking is a business, with focus on the opening weekend, the trailer, and the movie's market. Costume designers get no compensation in connection with the merchandising of their work. There is absolutely no participation for any member of the creative team that so often makes a significant contribution to the movie's success. It is time that this changed.

One of my most important tools as a costume designer is the ability to get along with people, the ability to collaborate with the director and crew. You need to lead and inspire and cajole whoever is working with you. Another quality that I rely on is my memory. I often use little pieces of information that I've tucked away in my mind. It might be one detail that I saw on the back of a magazine, or it might be a whole vista that I noticed. I think that this process is subliminal for the most part. I like to continually put myself in new situations and have new experiences, because you never know when the information and the imagery you absorb could be useful later.

Lastly, a costume designer needs to have inexhaustible stamina and energy. I know I couldn't design a movie like **The Last Emperor** again. I could not meet the physical demands; I could not cope with the stress, the pressure, the

1

2

3

logistics, the geography, and the hours. It's not only the industry that's changed, I've changed.

I would say don't go into this field unless you're really passionate about it. The indispensable quality you must have to be a costume designer is a love of human beings. You've got to love actors, and you've got to love your crew, and you've got to love your director. Even when you do, you've got to somehow create your own happiness and satisfaction through all of it. Otherwise, you will go crazy.

4

(1–4) **Restoration**: Director Michael Hoffman's theatrical vision of the life of King Charles II following the Restoration in 17th-century England. Acheson's lush, flamboyant costumes combine with Eugenio Zanetti's production design to create a debauched air to the film, and were actually created on very little money, using cheaper materials like stuffing flock and rubberized lace. Acheson gives credit to leads Robert Downey Jnr and Sam Neill: "They weren't dominated by the sets and costumes." Both Zanetti and Acheson won Oscars for their work on the film.

Milena Canonero's first film costume design was for Stanley Kubrick's classic **A Clockwork Orange** (1971). Canonero's close collaboration with Kubrick continued with the extraordinary **Barry Lyndon** (1975), and she won her first Academy Award for costume design on this, her second film. Known in the industry for both her talent and her attention to detail, Canonero has worked with some of the most highly

milena canonero 25

respected and innovative directors of the last 30 years. She designed lush costumes for Francis Ford Coppola's **The Cotton Club** (1984), **Tucker: The Man and His Dream** (1988, nominated for an Academy Award for Costume Design), and **The Godfather, Part III** (1990). Her sensitive period and ethnic costumes for Sydney Pollack's **Out of Africa** (1985) were nominated for an Oscar, and she worked with Louis Malle on **Damage** (1992), Norman Jewison on **Only You** (1994), and Roman Polanski on **Death and the Maiden** (1994). The evocative costumes for Hugh Hudson's **Chariots of Fire** (1981) brought Canonero her second Academy Award, and she was nominated for her groundbreaking designs for **Dick Tracy** (Warren Beatty, 1990), **Titus** (Julie Taymor, 1999), and **The Affair of the Necklace** (Charles Shyer, 2001). She was honored for her contribution to the art of filmmaking in 2001 with The Costume Designers Guild Award. Although some of her costumes have resulted in fashion trends, Canonero doesn't think about fashion when she works. Her feeling is that, "If you capture a trend, it's already there. It's in the air."

I was studying in London, and had become friendly with Stanley and Christiane Kubrick when he asked me to work with him on **A Clockwork Orange**. Of course, I knew the book, and I was very taken with it, like many young people at the time. I was passionate about filmmaking, but I had no real experience. Stanley reassured me that he wanted a fresh approach, which is what somebody just beginning could bring him. He introduced me to moviemaking by immersing me in the creative process. He said, "First, you help the production designer, John Barry. I want you to learn how to scout locations. In the meantime, think about the movie. I don't want science fiction. It's more ambiguous. It's now. It's tomorrow. We'll talk about the costumes once you have a sense of what I'm trying to achieve." Stanley gave me his Nikon camera with a wide-angle lens, and off I went, taking hundreds of photos all over London. I was inspired for the film's costumes by the skinheads on the street in London. I devised a stylized look for our "Droogs," and designed all the costumes to create a sort of surreal imagery. Stanley taught me to always keep the concept of the movie in my mind. He also taught me to start with the head, and immediately put me in charge of supervising and giving input on the makeup and hair. He said, "Movies are mostly close-ups." It was a wonderful experience, and Stanley's approach to moviemaking marked me forever.

Stanley asked me to design the costumes for **Barry Lyndon** with Ulla-Britt Søderlund. I found the clothing that was built in costume houses unacceptable; stiff with big padded shoulders. Stanley decided to take a different approach, and we rented an unused aircraft factory to use as a costume-making workshop. We hired students from the best costume design schools. I tracked down a priceless tailor and cutter and a costume maker, Gary and Yvonne Dahms. And a very shy milliner. We had started a trend. We were able to make clothing with original patterns from the Victoria & Albert Museum in London. We bought clothes and original laces and trimmings from collectors and auctions, and we made nearly all of the stock for the film. We were aiming for an elegant, subliminal vision of the 18th century. As on **A Clockwork Orange** I worked with Barbara Daly for makeup and Leonard for the hairstyles, and Stanley, once he saw we were on the right track, left us on our own and gratified us with his trust.

I like to steep myself in an era, collecting books and references. For **Barry Lyndon**, I traveled all over Europe, visiting libraries, and buying books on all aspects of the period—not just the costumes. It's useful for me to go as deeply as possible into the time being portrayed. It helps me to find an angle on the period, rather than create an academic look. For contemporary films, I look at photographs and paintings for inspiration. You never know what will generate a great idea.

I had only about two and a half months to prepare for **Out of Africa**; I went to Denmark to visit the writer Isak Dinesen's family, and examine her family archives. I wanted to achieve Dinesen's look, but also design clothes that were right for Sydney Pollock's movie. The British Museum and the London Library were great sources for research. But **Out of Africa** turned out, for many logistical reasons, to be one of the most difficult movies to work on, but it served me as a great lesson. So much happened on that movie; some happy, some sad, and some even grotesque, like when we discovered that our wardrobe driver and our guards were giving away some of our best costumes to local prostitutes in lieu of favors. We also had to be creative in the research for **Midnight Express**. We weren't painting a very flattering portrait of the Turkish criminal justice system, and I couldn't call the police about what their uniforms were made of. I played "tourist" and took lots of photographs for costume and production design purposes (including photos of the airport, which was under strict surveillance).

I had a ball working with Francis Ford Coppola on **The Cotton Club**. Francis came on board as the writer and director at a very late stage in the process. Dick Sylbert, the production designer who had suggested me to Bob Evans (the producer), and I had been brought in several months before, and Francis accepted us. I loved working with him right away. I had done a great deal of research, but because we had no definitive script, we were speculating on the period and situations. I had collected a lot of real clothes from the 1920s and '30s. There is something about the actual clothing of a period that you cannot easily recreate. Hats, in particular, are precious. The craftsmanship of making hats is very difficult, and you can rarely find someone with the talent to do it

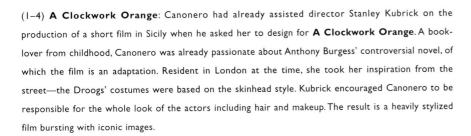

(1–4) **A Clockwork Orange**: Canonero had already assisted director Stanley Kubrick on the production of a short film in Sicily when he asked her to design for **A Clockwork Orange**. A book-lover from childhood, Canonero was already passionate about Anthony Burgess' controversial novel, of which the film is an adaptation. Resident in London at the time, she took her inspiration from the street—the Droogs' costumes were based on the skinhead style. Kubrick encouraged Canonero to be responsible for the whole look of the actors including hair and makeup. The result is a heavily stylized film bursting with iconic images.

(1–4) **The Cotton Club**: For the first three months of production on this film, there was no director. During this time, Canonero built up a collection of around 50 files on the history of New York and Harlem in the 1920s and '30s—the period in which the film is set. She visited museums and libraries in New York, buying hundreds of books, and photocopying styles she liked in order to create her research library. Although Canonero bought stock of authentic period clothes, the costumes for the musical numbers were all made, as they had to be able to move appropriately. Her work won her a British Academy of Film and Television Arts award.

3

1

2

4

properly today. I collect hats, and I use them in combination with the costumes I design whenever possible. Because the story was being created as we went along, we had to be ready for anything. I also loved working with all those dancers and singers. So much went on both behind the scenes and in front of the camera. **The Cotton Club** was one of my most exciting experiences.

For **Tucker**, magazine advertisements and early movies served as a great inspiration. Francis was aiming to recreate the look of the American films of the 1930s and '40s, while presenting a fresh view of Tucker's story. By the time I designed the costumes for **Tucker**, I felt like a part of Francis' family.

I like to work closely with the production designer and cinematographer. I like to have a continuous exchange so that our collaboration results in a complete whole. Vittorio Storaro, the cinematographer on **Tucker**, always sends a memo describing his approach for the movie to the director, production designer, and costume designer. It is not a "bible," but a very useful guideline on how he plans to light the movie. Other cinematographers I've worked with have been helpful, but Storaro is remarkably articulate. We worked together again on **Dick Tracy**. This movie was inspired by the comic strip, which used only five colors. I was interested in expanding those confines and in designing something more intriguing than that. Warren Beatty agreed to expand the palette to include ten colors, and Vittorio assured us that he would use a special process to enrich them. I determined the ten colors for the movie. Dick Sylbert, the production designer, asked me for swatches of the fabrics, and used only those colors on the sets. Using light, and giving the colors

different hues, Vittorio created a magical effect. In this respect, **Dick Tracy** beautifully integrated production design, cinematography, and costume design.

When I was prepping for **Out of Africa** in London, I often spoke on the phone with Stephen Grimes, the production designer. He arrived on location in Kenya before I did, and he would describe for me what he was doing in each scene. He sent me color copies of his sketches and paintings, so I could see the colors he was using, and his approach to the movie. I was using a lot of white and cream, and as Stephen had worked on **The African Queen** years earlier, we joked that we were making "The African Cream". Sydney Pollack saw none of my work until we did the screen tests. He was concerned that my color palette was dull but, with Stephen's unflagging support, was convinced that it would work.

I like to create costumes that are in harmony with the director's concept of the movie. Regardless of his method, I must understand what he is aiming for. When I was working on **The Godfather, Part III**, Francis spoke of an "operatic" movie with a Shakespearean undertone. I used Francis' cues, along with Gordon Willis' lighting (very much in the style of the Italian painter Caravaggio), and Dean Tavoularis' rich and timeless sets, to reinterpret the clothing of the 1980s. Some directors like to be involved in all the stages of preparation, others only the initial stage. Others still only want to see the costumes at the very final stage, ready to shoot on the set. Like with Francis. He says it is exciting to see what I have prepared. The actor's personality, vision, physical aspect— even his availability—are all going to become elements in the concept and the realization of his or her costumes, hopefully a good result will come out of it.

1

2

3

4

5

6

(1–6) **Barry Lyndon**: The recipient of an Oscar and a BAFTA nomination, the costumes for 18th-century period film **Barry Lyndon** were impeccably researched. Kubrick, the director, sent Canonero to Italy, Germany, France, Scotland and Ireland in the quest for source material: "Sometimes he said to me, 'If the books are not too expensive, buy two or three copies; then we can cut the costumes out, instead of making photocopies'." Canonero co-designed with Ulla-Britt Søderlund, dividing the designing and making between them respectively, and they had up to 50 other people working on the costumes. Canonero dismissed the stock she was presented at costume houses, saying it was, "Fake, phoney; it didn't move. It looked like tapestry, sofas." Consequently, few of the costumes were rented, and the rest were made.

It was simply delightful working with Al Pacino on **Dick Tracy**. For his role as Big Boy, Al improvised for me, with costumes and props, for several hours. I filmed him for more than two hours on my video camera while he invented the character right in front of us, and when he was done, I knew what his costume should be, and how to reshape his body with it. Some actors understand that the costume designer's work starts with the first hair on their head, and continues down to their toenails. Those who aren't afraid to take their image in new directions are the best to work with. For **The Godfather, Part III**, I designed Al Pacino's look on a computer. Francis liked the salt-and-pepper crewcut look with a receding hairline. Al, who had very thick, black hair, agreed to the arguably unattractive aging with no complaints. Hair and makeup, although they are not physically executed by costume designers, are critical elements of a character's plausibility and look. Therefore I always prepare research references and sketches for the hair and makeup team, and then meet and discuss the desired result together.

I am totally against making a separation between hair and makeup, and costumes. We do not start from the neck down. Like production designers with their set decorators, hair and makeup must come under the control of the costume designer.

In contemporary movies, often, our work is less obvious, but I try to find, beside the palette, something more satisfying. When I designed the costumes for Louis Malle's **Damage**, it was one of these experiences. He was very pleased that I was designing a look for the actors that was as elegant and stylish as was required, but that underlined the neuroses and the psychology of the characters. I hate it when people think that a contemporary movie is not really costume-designed because so much is bought. It is like saying that a production designer does not art direct because the movie is shot on existing locations. I always insist on making a certain amount of costumes in a modern movie but I do believe that selections and choices constitute designing a look.

Although we had very little preparation time for **Chariots of Fire**, we made most of the costumes for the men, and combined real period clothes and newly made ones for the women. Like **Out of Africa**, the costumes had a great impact on fashion. It was as though the fashion world was ready for the period style of each film, and the costumes just ignited something that was in the air. But I never think about fashion when I'm working on a movie. For me, the most important consideration is to work with interesting directors. Working with Julie Taymor on **Titus** was a particularly stimulating experience. She is truly a visionary, and she has a generous heart—rather than controlling… she inspires. She had more precise ideas than many other directors I've worked with. I elaborated on her ideas, took them further, sometimes I took a different tack altogether. It was difficult to crystallize a style for **Titus**; I used many sources to create a cohesive theme. It was like creating a very intellectual comic strip. We set up a workshop at Cinecittà, and we made everything in our workshop, including the stylized armor.

Barbet Shroeder, the director with whom I had worked on **Barfly**, asked me to production design and costume design **Single White Female**. Encouraged and well advised by my friend Dick Sylbert, I jumped into this wonderful new experience. The film was small and intimate enough to be able to coordinate the two duties, and I liked the creative experience of a continuous mental flow between the costumes

1

3

4

(1–4) **Out of Africa**: Director Sydney Pollack did not see a single preparatory sketch for the costumes of his 1985 safari love story. There was a short preparation time—only two-and-a-half months—and director and costume designer were not in the same country. But Pollack trusted Canonero, who benefited from a close working relationship with the production designer on the film, Stephen Grimes. He would send her Xeroxes of his sketches and paintings so they could use colors from the same palette. Sketch of safari outfit and still of Meryl Streep (1, 3), who played Karen Blixen-Finecke, the Danish plantation farmer, who falls in love with Denys Finch Hatton (Robert Reford, 2, 4).

2

(1–4) **Dick Tracy**: Warren Beatty as the eponymous comic strip hero (1–2), Madonna as singer Breathless Mahoney (3), and sketch of costume for Tracy (4). It is always a challenge to recreate the visual world of the cartoon comic strip on film. When Canonero was brought on to design the film, her first for director Warren Beatty, she was concerned that the strip artwork consisted of only five colors, and expressed her worries to Beatty, saying, "I want it to have a richness." Ten colors were eventually used. Canonero would give her fabric swatches to production designer Dick Sylbert, who repeated exactly the same colors on the set. She feels it is one of the most important films of her career, as production design, cinematography, and costume design were perfectly integrated.

4

and the art direction, and discovering how this brought me closer to the cinematographer and, of course, the director.

I recently designed costumes for Steven Soderbergh's **Solaris**, a sort of metaphysical, science-fiction, love story. Steven wanted to make it very real and enigmatic, so my work had to be very understated and minimalist. My goal was for the costumes to be almost unnoticeable and, I hope, timeless, especially as this followed **The Affair of the Necklace**, directed by Charles Shyer, which was so rich and plush. Steven Soderbergh served as the director, writer, cinematographer, and producer and editor, and his first assistant, Greg Jacobs, was also the executive producer, so questions were answered quite easily. Their sense of humor and style made the experience a pleasure for everyone.

I always get more involved in the movie itself than the costumes I design. I enjoy the artistic and intellectual challenge, even though I am never quite satisfied; but after all, movies, more than costumes, are my great passion.

The productions sponsored by local summer youth programs were Ruth Carter's introduction to dramatics. Although she planned to study special education, Carter's love of the theater drew her to a theater arts degree at Hampton University. With no program in costume design at the school, Carter taught herself the skill of telling stories with costumes while working on the college's plays. Carter was developing a

ruth carter

portfolio of costume sketches, when a chance meeting with director Spike Lee sparked a career in costume design for film. He was so impressed by her self-assurance, obvious talent, intelligence, and originality that he immediately involved her as a close collaborator on such films as **School Daze** (1988), and their partnership has continued to the present. Their joint projects include **Do the Right Thing** (1989), **Mo' Better Blues** (1990), **Jungle Fever** (1991), the inspired biopic **Malcolm X** (1992), for which she was nominated for an Academy Award, **Crooklyn** (1994), **Clockers** (1995), **Summer of Sam** (1999), and **Bamboozled** (2000). Carter's credits also include three films with John Singleton, **Rosewood** (1997), **Shaft** (2000), and **Baby Boy** (2001). Her Academy Award-nominated costumes for Steven Spielberg's **Amistad** (1997) exemplify her large range as a designer and her ability to design in every single genre, whether the story demands delicate, tea-stained, period dresses, or believable, street-savvy, modern costumes.

I was 25 years old, from Massachusetts, cruising in my Volkswagen Rabbit down the 5 freeway, heading west towards Los Angeles after an internship at the Santa Fe Opera. As I got closer to the city, the California Highway Patrol performed their "round robin" right in front of me. It felt like I had a police escort—a welcoming committee. Even though they were just slowing the traffic down, I was thrilled.

That year I was hired as a dresser for the Los Angeles Theater Center's first season. I soon managed to talk my way into the position of costume shop foreman, something I was truly not cut out for. I used my resources there to freelance design for a local dance troupe (with the emphasis on free). It was through this activity that I met Spike Lee, who attended a show I'd designed. Although he was then an up-and-coming filmmaker, he wasn't very well known... yet. Spike was with a friend who introduced us, and we began to talk about how to gain experience in film, something I hadn't even considered doing. A while later he invited me to a prescreening of his film **She's Gotta Have It**. I was working every day in the theater, but

remember thinking, "Let me at it! I can do that!"

Not long after, Spike asked me to design the costumes for **School Daze**. I immediately quit my theater job and started breaking down the script. No contract, no negotiation, just passion. I drew every character in the script, including those who only had one line. I was insane. I left L.A. for a short time, to solicit guidance from my brother, a visual artist in New Hampshire. This experience changed my life by showing me how to artistically organize and communicate many ideas. I treated it like a job, and I worked very hard.

When I was done with my drawings, I packed up everything into an enormous portfolio case and took it with me on the bus to New York. Spike gave me really complicated directions on how to take two subway trains to his apartment in Brooklyn. I stumbled down the steps and through the turnstiles with this monster case. I was exhausted by the time I arrived at his door. I laid everything out for him and made him sign each drawing if he approved. I still have those drawings with the little "Spike Lee" signature in the corner. Spike is a man of few words, and although he never said it directly to me, I knew he was happy with what I'd done.

I returned to L.A. and waited to get on the books. After about six months, I was finally put on the payroll for **School Daze**. Even though I'd done so much design preparation, I wasn't ready for the process of fitting so many actors and being pulled in every direction by so many people. The politics of a movie set was one thing I hadn't bargained for. But in the end, I was pleased. After this, I continued working with Spike Lee,

and I designed costumes for eight more of his pictures. This proved in the long run to be a much greater "thank you" than a few words.

One major difference between designing costumes for the stage and designing costumes for the screen is the aesthetic distance. As a theater intern, I painted fabrics and made bucket boots. I created a lot of larger-than-life costumes and accessories, because fine details are lost on the stage. For the theater, you can take a modern shoe and adapt it for another period, and from a distance it looks great. I think distance, in combination with lighting, creates a big difference between the two media. In film, the camera is your eye, and it is much more intimate.

The success of **She's Gotta Have It** generated an interest in Hollywood to produce and promote the work of black filmmakers. A surge of films by the black community were made, and my next movie was a spoof called **I'm Gonna Git You Sucka** with Keenen Wayans. This was a very different experience for me from **School Daze**. Unlike Spike, Keenen would tell you, all the time, how much he liked the work you were doing. Sometimes I drew sketches through the night in my little apartment. The movie was a spoof and was fun to do.

Halfway through the preparation for **I'm Gonna Git You Sucka**, Spike asked me to design the costumes for **Do the Right Thing**. I knew I couldn't work on both films simultaneously (on opposite coasts, no less), but I tried it anyway. I remember standing on 34th Street and talking on two pay phones at once—New York in one hand, and L.A. in the other. It was crazy, but from a creative perspective it worked out well. I have great memories of working on **Do the**

2

3

4

(1–5) **Malcolm X**: Spike Lee directed this biopic of the influential figure Malcolm X. This was the sixth film that Carter had designed with Lee, and the fact of her existing relationship with the director allowed her complete artistic freedom. The film garnered her an Oscar nomination. The costumes had to reflect the powerful change that Malcolm X's character went through—from small-time gangster Malcolm Little (4–5) to devout Sunni Muslim (1–3).

5

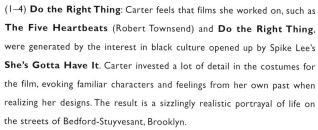

(1–4) **Do the Right Thing**: Carter feels that films she worked on, such as **The Five Heartbeats** (Robert Townsend) and **Do the Right Thing**, were generated by the interest in black culture opened up by Spike Lee's **She's Gotta Have It**. Carter invested a lot of detail in the costumes for the film, evoking familiar characters and feelings from her own past when realizing her designs. The result is a sizzlingly realistic portrayal of life on the streets of Bedford-Stuyvesant, Brooklyn.

Right Thing. The first thing that struck me about the script was the offbeat but familiar names of the characters: Mookie, Mother Sister, Da Mayor. They were immediately alive for me. I didn't grow up in New York and I didn't have the type of brownstone community experience depicted in the film, but I could relate so well to the story—the hot, hot day, the urban environment, the collisions of love and hate. I felt as though I was really expressing myself in my designs for the film; the costumes projected an image of what I saw in my heart. When I dressed Mother Sister and Da Mayor, I thought about the specific things that I love about people, and how their flaws and their endearing qualities can be subtly portrayed by the way they look. Even though Spike is not communicative, he brings you into his life; you become a part of his world. It's Spike's experience, and I learned so much about New York and New Yorkers from this collaboration.

Two favorite projects stand out in my mind. **Malcolm X** was a joy for me because I felt truly, artistically free. I already had a relationship with Spike, and he left me on my own. I had four floors of costumes, and I could do whatever I felt was right. Similarly, Steven Spielberg gave me a great deal of freedom to make my own decisions on **Amistad**. These two films were quite exceptional in that way, and, oddly enough, these were the two movies for which I was nominated for an Oscar.

I was so excited when Steven asked me to meet with him about **Amistad**. It was an informal discussion, and he offered me the job right there. The supervisor I chose had worked with Steven before, and she advised me that I could do whatever I needed to to get the job done. I took her at her word, and traveled to Rome to pull at Tirelli Costumes, and to London to pull from the costume houses there. This was a

huge difference from my past experience. When I got back I did a presentation for Steven in the form of a collage that included about 30 plates of research, and some rough and some final sketches. He loved it. Throughout the whole process he provided a lot of positive feedback. Sometimes on the set of **Amistad**, I felt like the scenes were actually happening. When a scene works that well, it is like an out-of-body experience. That, to me, is how I gauge my success.

Rick Carter was the production designer on **Amistad** and I worked very closely with him. He was always dropping by. I had the same type of experience with Steve Altman on **What's Love Got to Do With It?** He kept me in the loop, even though our paths did not cross often. He would do things like leave me a brochure from the 1960s in my mailbox, with a note saying, "Look at these colors." It's really helpful when a production designer does things like that—specifics are more useful than general, conceptual discussions about the design of a film.

To me, the essence of designing costumes for movies is to decide what your style is going to be for that particular picture, and not wavering from your decision. You have to stay focused on keeping the costumes consistent with the look of the film, the look that you will generate with the director and the production designer. Is it whimsical, romantic, realistic? Sometimes I don't get a sense of the style until I do a good deal of research.

My most challenging project was **What's Love Got to Do With It?** Of course, the movie was based on the life of Tina Turner, a real person who has created a distinctive image. I repeatedly heard the complaint that Angela Bassett, who

1

2

3

4

5

6

7

8

(1–10) **Amistad**: This film of a slave mutiny, set in 1839, required Carter to thoroughly research costumes of the period. Djimon Hounsou plays the slave Cinque (6–8), who is on board a slave ship bound for the New World. The passengers mutiny, are captured, and a court case ensues to determine their fate. Anthony Hopkins is the former President of the United States, who argues the slaves' case (3), and Morgan Freeman plays an abolitionist (4–5). Sketches of costumes for the principal characters, signed by the actors (1–2, 9, 10). Steven Spielberg, who directed the film, already admired Carter's work on **Malcolm X** and **Do The Right Thing**, and encouraged her to have complete artistic freedom.

43

9

"The early Baldwin" my favorite "get up" of the film.

10

1

2

costume design

(1–4) **Jungle Fever**: Carter has a strong grounding in designing for period pieces, due to her extensive work in theater. "Modern pieces are harder for me, because I don't come from a fashion background, so I kind of learned as I went." **Jungle Fever**, another Spike Lee film, required highly contemporary costumes for its portrayal of mixed-race relationships in Manhattan. Wesley Snipes is Flipper Purify (1, 3–4), the married architect who falls for a white girl.

3

played Tina Turner, didn't look like the singer. The Disney executives couldn't seem to get beyond the fact that the two people just don't physically resemble one another, and they were convinced that the problem was caused by the costumes. I was frustrated and upset; I wanted to quit the picture. The executives offered to get me some assistance, and gave me a list of designers I could consult with. One of them was Ellen Mirojnick, and I gave her a call. She was a wonderful support. She said, "What the hell am I going to tell you about dressing a black woman?" I got an earful. I said, "She's the perfect person to bring in." So Ellen showed up, and she looked through the photographs of the costumes. She told me she didn't see anything wrong with what I had done. In the end, I was proud of the costumes in that movie, but I would never want to have another experience like that.

For me, designing costumes has changed over the years. When I first started, I experienced a sort of pure collaboration with the director. Then I was designing costumes for smaller, independent pictures, and there wasn't much in the way of studio oversight. Now I'm working on studio pictures with bigger stars, who have different requirements from less well-known performers. Directors will occasionally need studio approval on designs. Overall, the project becomes less and less my own.

Someone once told me that my work is distinctive in my use of red and geometric forms. To me, my honesty, my take on life, how true I can make something, are the elements that distinguish my work. I think it's seen best in the period pieces, because that is where my training is and where I am most comfortable expressing myself. Modern pieces are harder because I don't have a background in fashion. For contemporary films, I've had to use a less structured approach than on period pieces, and learn as I go.

A costume designer should be interested in people and be able to interpret the reasons why they make the clothing choices that they do. There is a certain psychology of costume design—you are helping the actor to create the character and to enhance the story. I look for something to completely draw me into the movie, regardless of when it is set or what it is about. Becoming engulfed in a new world is what I enjoy most about designing costumes for films.

I feel a spiritual connection with the African-American experience in this country, but I don't feel limited to designing for films featuring African-Americans only. My designs are honest because I'm a person of the people.

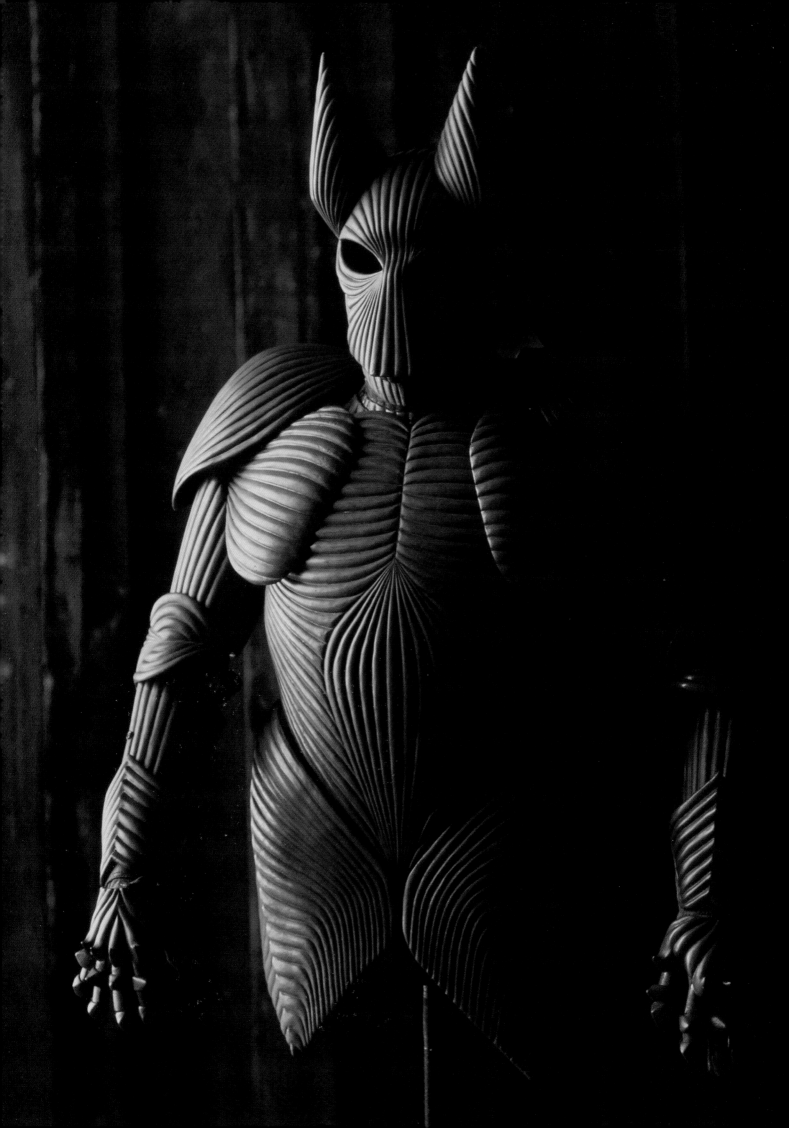

A graduate of Tokyo National University of Fine Arts & Music, Ishioka has won international acclaim as a graphic designer and art director, a set and costume designer for film and theater, and as a director of television commercials and music videos. Her long list of accolades includes an Academy Award for costume design for Francis Ford Coppola's **Bram Stoker's Dracula** (1992), the Award for Artistic

eiko ishioka

Contribution at the Cannes Film Festival for her production design of **Mishima: A Life in Four Chapters** (Paul Schrader, 1985) for which she also designed the costumes, Tony nominations for the sets and costumes of the play *Madame Butterfly* and a Grammy Award for the artwork of Miles Davis' *Tutu* album. Her more recent activities include the costume design for a production of Richard Wagner's epic opera *Der Ring des Nibelungen* at the Netherlands Opera House, costume design for **The Cell** (Tarsem Singh, 2000), direction of a music video for the singer Björk, and competitive sportswear design for the 2002 Winter Olympics. In the spring of 2002, she completed the costume design for the new Cirque du Soleil production *Varekai*. Ishioka is a laureate of the New York Art Director's Club Hall of Fame, and her work is included in the Museum of Modern Art in New York, among others. She has published two retrospectives of her work, *Eiko by Eiko* in 1983, and *Eiko on Stage* in 2000. Ishioka was born in Tokyo and resides in New York City.

interview

My father was a graphic designer and I was influenced by him in many ways. I studied all aspects of design in college, and after I graduated, began my career in graphic design. Before long, I became an art director in the advertising world, directing entire corporate advertising campaigns and, in particular, directed many television commercials and advertising posters.

At about that time, I was approached by Japanese theater directors and contemporary dance choreographers who were intrigued by my freedom of expression in various mediums, and wanted to collaborate with me in new and experimental ways. My first foray into costume design was through an avant-garde Japanese choreographer who wanted me to do the sets and costumes for a stage production based on Stravinsky's *A Soldier's Tale*. Soon after that, I was approached by a theater director about doing the sets and costumes for Shakespeare's *Hamlet*. At the time, I did not consider these jobs to be the new trajectory of my career; I took them on more as an occasional change of pace from my other design work. It was much later in my career that set and costume design became something I really focused on.

I learned the foundations and theory of design at university, but I have never studied costume design per se. When I am designing costumes, I work with technical associates in much the way that architects work with engineers. When approached to work on a production, it is almost always because the people in charge are looking for a very bold, unconventional visual statement. I rarely rely on traditional means to realize my designs. The craftspeople on the production and I work together to find new directions and new methods, and often discover new techniques in the process, which is always very exciting.

I have learned the meaning of creative expression from a wide variety of people working in an even wider variety of creative mediums. Fashion designer Issey Miyake has been a friend since I began my professional career, and we have inspired and influenced each other creatively over the years. I've known Leni Riefenstahl for 25 years, having produced, directed, and designed two major exhibits of her work at museums in Tokyo. Irving Penn's photography has influenced my work from the very beginning, and the many precious stories he has shared with me have inspired and nurtured me throughout my career. No matter what creative field they come from, they all responded unanimously to my query: "What is your most important trait as a creator?" The answer: Discipline. Any creator who does not nurture a sense of self-discipline cannot continue doing good and innovative work for long.

From the beginning of my career, I have drawn creative inspiration not just from professional artists, but equally from ordinary people in everyday encounters, nature, travel, and history. I've traveled extensively and my experiences have left me with the sense that the whole world is "my studio" and an endless source of motifs for my work. My parents, who exposed me to many of life's great treasures and nurtured my creative spirit, have been a tremendous source of inspiration.

Before starting a project, I have to fully understand its content. That involves analyzing and breaking down the script, and talking with the director, producers, actors and other crew members to really grasp the important points of the film. My first task once I begin designing is to face a blank piece of paper and let my mind wander limitlessly through what I want to express. When I have settled upon a direction, I will gather resources as necessary. Research materials are always merely hints for further developing ideas. Period pieces and historical stories naturally require thorough research and an understanding of the visual world of the specific time period, but here too, I will "recook" the historical cues to create my own tableau.

After first listening to a director's vision about the characters and then analyzing and digesting this information, I begin building ideas through a process of trial and error. In other words, I try to avoid handing a director detailed material right away, making my initial idea sketches a way of communicating my general concept rather than presenting a polished design. Sometimes I make several intermediary sketches before producing detailed, large-scale drawings; other times I jump directly from the initial sketch to the final design. As you might expect, final drawings need to depict every single detail clearly defined and in color, and—above all—convey a unique and compelling design. My sketching

1

2

(1–3) **Mishima: A Life in Four Chapters**: American writer/director Paul Schrader asked Ishioka, who had previously worked mostly on set design for theater productions, to come on board **Mishima** as production designer. The general's meeting in the *Runaway Horses* segment of the film (1). The retro set design for the *Kyoko's House* segment (2) was inspired by 1950s American kitsch. Ishioka also drew upon iconic traditional Japanese imagery for set and costume design, as in this scene of the *kendo dojo* (sword-fighting hall) in the *Runaway Horses* segment (3).

eiko ishioka

3

1

2

(1–7) **The Cell**: The costume sketch and finished view of the stunning, bird-inspired dress worn by Dr Deane (Jennifer Lopez) in the dreamlike virtual reality sequence at the film's opening (1–2). The final drawing (3), close-up (6), and still (7) of the rubbery body suits, conjuring up images of the muscles in the human body, that Dr Deane and her colleagues wear when entering their patients' mindscapes. Dr Deane's armor, including a metal visor and rubber neckbrace, for her battle with Carl Stargher (Vincent D'Onofrio) in the film's finale (4–5).

3

4

5

6

7

style varies from project to project. Often I leave nothing to the imagination, giving the sketches a great deal of graphic detail. My final drawings serve as a vital communication tool for the entire production staff—everyone from producers, the director, actors, director of photography, visual and special effects team, to the costume assistants and craftspeople who execute my designs. The drawings, depicting details of how the head, face, and various body parts should look, act as storyboards and allow a smooth collaboration with makeup artists, hair stylists and others.

Mishima was my first collaboration with a film director from the West, and my first foray into American film. When Paul Schrader asked me to come on as production designer, I asked him, "Aren't you nervous about my lack of experience in film?" He replied, "We can resolve that by teaming you up with a great technical supervisor. What I need from you is a totally new and revolutionary design concept for the film." The Japanese film world is rife with male chauvinism, and the fact that the American producers and director had chosen a Japanese woman, an outsider to the film industry, as production designer, was not appreciated, to say the least, by the local crew members. I faced harassment and resentment on a daily basis, and this made working on the film a trying experience. Thanks to the undying support of the American filmmakers, however, I was able to realize my designs and went on to win an award at the Cannes Film Festival.

Paul Schrader asked me to join his production team on **Mishima** because, as he told me, "I want to experiment visually with the film, and I thought it would be better to work with an outsider to the industry." Indeed, **Mishima** was my first major film project and it was certainly a trial by fire.

Thanks to the open-mindedness and collaborative spirit of Schrader, however, it was to be the first of several great film projects for me. Schrader told me during the filming of **Mishima** that it is the collective brain power of the director, cinematographer, and production designer that makes for a great film. He is one of the few directors who puts a great deal of emphasis on the visual elements of a film, and accordingly, expects the designer to carry a full third of the weight.

He wanted to create a theatrical set-within-a-set for the sections of the film that would symbolize Mishima's life through three of his novels. Of the three novels represented in the film, the most difficult to visualize was the kitschy Japanese version of 1950s American culture of *Kyoko's House*. Though in general he is not the type of director to painstakingly explain his visual direction, Schrader had specific ideas for the set of *Kyoko's House*, and for the first time in my career I was asked to deliberately design something in bad taste. I rose to the challenge, however, and was rather pleased when some well-known American architects commented to me that my "bad-taste design" was quite fresh!

Up until Francis Ford Coppola asked me to do the costumes for **Bram Stoker's Dracula**, I had always thought of costume design as constricting: the designer interprets the director's vision, not his or her own, and most of the designer's time is spent doing period research and technical drawings. I know that's important work, but it's not terribly creative. I thought that the costume designer's role was essentially passive, and passive is definitely not my style. Another of the reservations I had about costume designing predates my **Dracula** experience, and unfortunately it persists. Both

1

4

2

3

(1–4) **The Cell**: In the film, Vincent D'Onofrio plays Carl Stargher, a notorious serial killer in a coma. Jennifer Lopez' psychotherapist character Catherine Deane has to travel inside his mind to find the location of his most recent victim. Within his inner mindscape, Stargher is an all-powerful king of a dark underworld, and Ishioka designed several fantastic costumes for this role, of which these are sketches and stills.

(1–5) **Bram Stoker's Dracula**: Ishioka won an Oscar for her costume design on Francis Ford Coppola's haunting film, starring Winona Ryder, Keanu Reeves, Sadie Frost and Gary Oldman, who played Dracula. Sketch and still of the Elizabethan-inspired wedding dress worn by Lucy (Frost) as a vampire (1, 3), and still, sketch and close-up of Elisabetta's (Ryder) sumptuous red gown (2, 4–5).

costume design and production design have always been seriously undervalued in film and theater.

So, when Coppola first approached me about doing **Bram Stoker's Dracula** in early 1990, I said, "Why are you asking me? I'm not a costume designer." The fact that I could say this to Coppola shows how much I resisted the idea, since I have always really admired him and his work. What finally hooked me was the fact that Coppola wanted the costumes to be the sets of the film. He wanted to use volumes of fabrics of various textures, colors, and patterns to create a rich tapestry that evoked the Victorian era and Dracula's world.

In Coppola's mind, the sets would act as the background for the interplay of light and shadow, conveying more of a mood than a tangible image. The costumes, on the other hand, would be an authoritative and defining presence, giving shape to each of the characters in the film. It was Coppola's unique vision and the freedom he accorded me to help realize it that ultimately made the costumes so successful in the film.

I sketched my first ideas for **The Cell** as line drawings off the top of my head—over 60 in all—while working on the costumes for the opera *Der Ring des Nibelungen* in my hotel room overlooking the canals of Amsterdam, and faxed them to director Tarsem Singh in L.A. I had prepared five or six different variations for each of the costume ideas in black line drawings; Tarsem responded unequivocally about some, and hesitated about others. At the time, Tarsem had to make some crucial decisions about the script, casting, and set, so he was wavering a bit on the final direction he would go with the characters. Reaching quick decisions about costuming, therefore, was also difficult.

On previous projects, Tarsem had given the designers very specific ideas about his vision. When he told me, "Make something just like this," showing me a costume idea culled from a magazine clipping or some reference material, I was offended—and told him so. Costume design, after all, is not simply a matter of copying a look one researches in historical reference books. Relying on a predetermined idea may make a director more comfortable and make it easier for them to envision what the final product will look like, but it also stifles a costume designer's creativity. Though he may have been unaccustomed to my working style at first, Tarsem adapted quickly and was extremely supportive.

I like designs that are innovative and monumental, not ones that simply explain the story or roles to the audience and end there. I want to create something else—to stir up the imagination of the audience, stimulate their eyes, and move their spirits. My work will be a great success if the audience says "Wow!" over and over again. Of course, my aim is not to stand out for the sake of standing out. The costumes should serve to strengthen and enliven the overall visual vocabulary of the film.

When I began my career—initially in graphic design—it was assumed that I would specialize in that one field, and that one field alone. Young designers were taught then, much as they are today, to hone their skills in a single discipline, and were rarely encouraged to explore the nearly limitless applications and interpretations of design. My path as a designer has been decidedly different. Through a combination of my own natural

eiko ishioka

1

2

costume design

3

(1–7) **Bram Stoker's Dracula**: These stills
show one of Dracula's gowns, inspired by the
erotic paintings of Austrian painter Gustav
Klimt (1–2). Ishioka designed this Dracula
family crest especially for the film (3), and it
adorns many of the sets and costumes
throughout (4). A finished sketch for the
costume of one of Dracula's drivers (5), and
the drawing and still of Dracula's armor in his
incarnation as Vlad the Impaler (6–7).

4

instincts and will, along with equal measures of serendipity, I have worked as an art director in advertising and corporate identity campaigns, as a set and costume designer for film, theater, and opera, as a director of television commercials and music videos, as a museum exhibit producer and designer, and much, much more. In what seems even to me to be a mysterious chain reaction, each design endeavor I take on is followed by one in a completely different and unrelated field. This provides for a great deal of creative freedom and constant artistic stimulation—the very air and water designers need to thrive. Design is a truly boundless frontier, and I consider myself blessed that I continue to be able to explore its many and varied wonders.

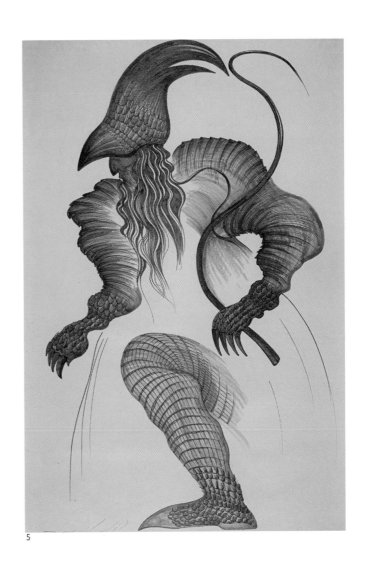

5

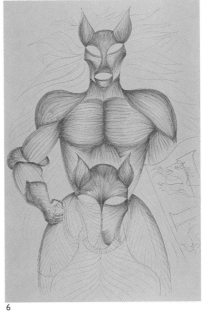

6

7

biography

Jeffrey Kurland is one of the few modern costume designers who has worked extensively with one director. Kurland's collaboration with Woody Allen (as costume designer on 15 films and assistant on three others) has given him the opportunity to design costumes from an almost instinctual perspective. This collaboration also provided a chance to design period pictures, such as **The Purple Rose of Cairo**

jeffrey kurland

(1985) and **Radio Days** (1987), for which Kurland won a British Academy of Film and Television Arts (BAFTA) award for Best Costume Design, as well as contemporary movies, including **Hannah and Her Sisters** (1986), **Crimes and Misdemeanors** (1989), and **Mighty Aphrodite** (1995). Kurland was nominated for an Academy Award for his glamorous contributions to Allen's hysterical 1920s comedy **Bullets Over Broadway** (1994). During his years in New York, Kurland also designed costumes for **This is My Life** (1992) and **Mixed Nuts** (1994), directed by Nora Ephron. More recently in Hollywood, Kurland has acquired a name for stunning contemporary costumes, and accurate characterizations, in such films as **Man on the Moon** (Milos Forman, 1999) and for P. J. Hogan's **My Best Friend's Wedding** (1997). The Costume Designers Guild honored Kurland with its award for Best Costume Design for a Contemporary Film for his remarkable, earthy transformation of Julia Roberts in Steven Soderbergh's **Erin Brockovich** (2000), and nominated him for the same award for Soderbergh's elegantly suited **Ocean's Eleven** (2001).

interview

My career developed after I finished college. I was working as an assistant costume designer to Patricia Zipprodt, a great stage designer, in New York. She arranged a job for me painting and distressing costumes for *La Gioconda* at the Metropolitan Opera. I had been working at Grace Costumes for a few days, doing delicate aging with paintbrushes and dyes. The head assistant saw that I was working altogether too slowly and came over with a great fat brush for me to use, and it might as well have been a roller. I was working on costumes for people at the back of the stage, which was 300 feet deep. I've long since learnt how designing for the stage differs from designing for film. The stage is big, and it is far away from the audience. Film happens in close-up most of the time. It's far more detailed, and you need to consider nuance and intricacies of character. You can design a stage piece and change it after the show's been running for two months, but you don't get that opportunity on a film. You must make your decisions quickly and then move on. If I designed for the stage now, I would approach it that way. I would never take the approach that I was creating something on a trial basis.

I started working on films in 1978 as an assistant costume designer, and my earliest experiences were on Woody Allen pictures. I had worked on **Stardust Memories, A Midsummer's Night Sex Comedy**, and **Zelig** as an assistant, and my first film credit as costume designer was for **Broadway Danny Rose**.

I worked for Woody for 16 years on 18 films. Costume designers consistently worked with one director during the age of the studio system, but this is something that just doesn't happen anymore. Our relationship was truly a collaboration. I would read the script and we'd talk about the story and the characters. As the writer, he never doubted what he wanted, but he trusted that I could riff off his ideas. First I would show him costume sketches, and then I would show him the actual clothes. For Woody's movies, we always made time to shoot wardrobe tests on every piece of principal wardrobe. We did them in front of a seamless backdrop, but I had to think about how the costume would look in context. I relied on Woody's instinct and my own. Most directors don't test wardrobe in this way because it costs too much money, and I'm fortunate that I got to have the experience. I learned a lot from it—both about what types of items work and how film stock affects wardrobe. Once we started filming, I would watch the dailies (the unedited picture and sound material from the previous day's shoot) and take notes. Sometimes I had only two or three days to make changes, but over time I knew intuitively what Woody would like.

Broadway Danny Rose is a black-and-white film. To design the movie's costumes, I didn't look at color, but rather at a fabric's hue and what its tone and patterns were. I used a Pancro glass, a device that basically eradicates color, so all you see are tones. In some patterns, the colors have the same tone and if you eliminate the color, you lose the pattern altogether. **Shadows and Fog** was also in black and white. It was set in the 1910s, and I had to design the costumes for people in a traveling circus. Concerned again with hue and tone, I was creating my own patterns using a variety of fabrics including velvet, silk, and brocade, which were quite bold and worked beautifully, providing the right sense of dimension. The film was a study in German expressionism, so I used color when I designed it. I thought it would have been done that way in Germany in the 1930s. In contrast, **The Purple Rose of Cairo** was a color film set in the same decade. Within the movie is a Hollywood musical from that era, and I designed the costumes in black-and-white fabrics, as Adrian might have done. The production designer, Stuart Wurtzel, and I decided that everything would be in tones of gray, black, and white, and it served our purposes well.

I always talk about color with a film's production designer. I'll get paint chips from him showing the interiors of sets and furniture and then choose my colors. I had a great relationship with Santo Loquasto, who was the production designer on many of Woody Allen's movies. We would visit each other's studios as our work progressed. The amount of give and take you have with a production designer is important.

As a designer, I plan ahead as much as possible, but I have to be resourceful too. In **Mighty Aphrodite**, a film I designed for Woody, Mira Sorvino played a hooker. I had her in skintight pants, and a little midriff top. She was all dressed, but when we arrived on the set, the soundman told me he would have to mike her. With this costume there was simply no place for a mike pack. So I had her remove the pads from

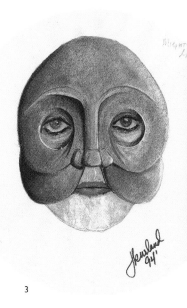

1

2

3

(1–5) **Mighty Aphrodite**: Starring Mira Sorvino in an Oscar-winning role (4), **Mighty Aphrodite** is a contemporary story, underscored with a traditional Greek chorus. Sketches of costumes for members of the chorus, which existed to link the modern storyline with that of the tragic Greek figure of Oedipus (1–3, 5).

jeffrey kurland

4

5

1

2

3

4

5

6

7

8

(1–9) **Bullets Over Broadway**: Dianne Wiest as the faded star, Helen Sinclair (1), who competes with the squawkish gangster moll Olive Neal, played by Jennifer Tilly (2), in a 1920s Broadway stage production. Sketches of two flamboyant costumes for the chorus girls (3–4), for Tracey Ullman's Eden Brent (5), Helen Sinclair (6), and Olive Neal (8). John Cusack plays the idealistic playwright, David Shayne (7), who falls in love with Helen (9). The film's costumes are a burst of color against a vibrant, roaring, 1920s New York.

jeffrey kurland

9

(1–5): **The Purple Rose of Cairo** features a film within a film, shot in black and white, while the rest of the film is shot in color. When researching the costumes, Kurland found costumes designed by other designers, including Adrian, that were made in tones of black and white, and he used these ideas to create costumes for the film within the film. All the set pieces for this section were also designed in tones of gray, black, and white. Still and sketch of Jeff Daniel's character, who plays Tom Baxter, the archeologist who steps out of the film into real life (1–2).

her brassiere, and had the mike pack placed into the brassiere next to her breast. Then her breasts were uneven, so I added another pad to the other side. I ran the wire up her arm and taped it to the bra strap and placed the mike in her hair—all of this in two minutes. Can you imagine, you're about to do your first scene in a Woody Allen movie, and the costume designer suddenly decides he has to rearrange your underwear? But Mira was a good sport.

This was not an unusual situation—a costume designer has to take it all with a grain of salt. He can collapse on the floor later, but he has to be cool in the moment. He can't make an actress crazy, and he can't let a director lose confidence in him. To my mind, this is why a costume designer's place is on the set. Movies are not made in fitting rooms. If he's not on the set, he's missed the boat, because as soon as he's gone, something's going to happen. And that costume is not going to look the way he wants it. Or the extra who is not "just perfect" is going to be placed right in front of the camera. I never leave the set. If I'm not there to worry about the costumes, no-one will do it for me. To be a successful costume designer, you have to be very flexible, but you also have to have a talent for chitchat. A fitting is often most successful if the actor is focused on the character, rather than the costume. So you want to be able to converse easily yet maintain your concentration. Every designer has their own way of doing it.

Designing costumes for a Woody Allen film is somewhat like designing for the stage, because Woody shoots everything in a master with very little coverage. You must get it right for the entire scene. He'll have the actors walk in and out of the room, and whatever happens to those costumes during all that action is what you're stuck with. So you can't put the mike pack in a place that's going to show if the actress turns around. When everyone made their entrance on the stage in **Bullets Over Broadway**, I had prepared for every movement they made. I couldn't take the risk, for example, that the fabric of Dianne Wiest's dress would catch on something and not fall properly when she stood up. I couldn't depend on coverage to pop in on a close-up or an over-the-shoulder shot. It didn't happen that way.

My first picture away from Woody was **My Best Friend's Wedding**, directed by P. J. Hogan. He and I needed time to get comfortable with one another. As a costume designer I had to gain his trust, just as I do with the actors. The director is the person I focus on. In the beginning, I don't think about the actors, I think about what the director wants and where he wants the picture to go. When you finally meet the actors, you don't have a lot of time to earn their confidence. In the film, Julia Roberts played a very stubborn "New Yorky" businesswoman. I had worked with her before on **Everyone Says I Love You**. Once Hogan and I had discussed her character, the two of them would meet, and then I would show her the sketches. We formed a wonderful triangle. As far as the designs were concerned, we were all on the same page from the beginning.

Later, I designed the costumes for **Erin Brockovich,** and I worked with Julia again. Erin Brockovich was obviously a real person, but we dressed the character to suit Julia's portrayal of her. It was fun because I actually got to speak with Ms Brockovich. I looked at her family photos and we talked about her style, and how it made her feel. It was the first time that my research was based on meeting with the real person being depicted, and it was a very special experience.

costume design

1

3

2

4

5

6

7

(1–7) **Erin Brockovich**: Kurland already had a working relationship with Julia Roberts, when she starred in her multi-award winning role as Erin Brockovich in Steven Soderbergh's eponymous film, based on the story of a real woman. The costumes in the film were, "very character-driven," not just for Roberts but for all the cast, including Albert Finney, who played people's lawyer Ed Masry (7). Kurland had to physically recreate Brockovich's body on Julia Roberts, and built her up using padded bras and other theater costume techniques. Drawing sketches (1–2, 4–5) were instrumental in forming a good understanding between designer, director, and actress.

1

(1–3) **Broadway Danny Rose**: This was shot entirely in black and white. When designing the costumes, rather than looking at their color, Kurland looked at fabrics through a special device called a Pancro glass (given to him by cinematographer Gordon Willis), which eradicates color and only shows tones. Kurland sees the film as having a "Runyonesque" quality to it, and cites it as one of his favorites. Commonly perceived as a period film, the film was designed as contemporary, and all costumes were manufactured. Mia Farrow as Tina Vitale, Woody Allen's girlfriend (1). Allen placed Kurland in charge of hair and makeup. On the Brooklyn housewife Tina Vitale's hair: "I said to the hairdresser, 'It should like she's actually wearing a Yorkshire Terrier on her head. That's the look.'"

2

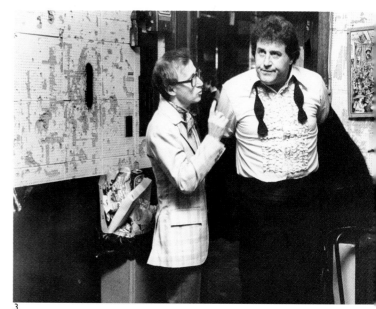

3

When I presented my drawings to Julia, she was looking at the artwork and colors, but more importantly she understood where we were going conceptually from her meetings with Steven. Working with directors in this way is extremely liberating. But it's also important in terms of getting the job done well. If an actor understands your perspective as an extension of that of the director, the project is really coming from one mind. I've been lucky because this has been my experience with Woody and Steven, but also with Neil Jordan on **In Dreams**. I did a period film with Milos Forman, **Man on the Moon**. Milos was extremely generous. He was great to work with because he had a clear concept of how he wanted to make his movie, and then he let me design within that concept. I knew my parameters before I even started. I want to give a director what I know is the right costume for a character, and I'll push for the costume that I think is correct. But I'll only push to a certain extent, because it's the costume designer's job to realize the director's concept. It has nothing to do with being subservient, it is simply that that is our job. It's the director's vision.

I approach contemporary and period pictures the same way. I read the script and then formulate my concept visually within that of the director. I discuss every single character, from the principal actor to the extras with the director. As a designer, I'm creating the costumes for every extra who passes in front of the camera. Not only what they are wearing, but its color, and how it will mix in with everything else that's on screen at the time. The term "extra" is unfortunate, because they're characters, too. They tell a story even as they're walking past. Sometimes, the people on screen who don't speak a word tell as much of the story as the principal characters who do.

Obviously, there's a difference in logistics between period and contemporary movies. When I'm designing costumes for a period film, I want to recreate the fabrics of the time. For contemporary pieces, I may do a good amount of purchasing, but I've never just bought an item and then put it on the screen. With character being created by a team—the writer, the actor, and the costume designer—one can't just go "find" some clothing and expect it to be right. Everything is filtered through a creative process. So even if an item is purchased, in the end, it too must be created, just as the writer creates the screenplay and the actor creates the character.

If I'm designing a contemporary film I try very hard not to date myself. I don't ever want a viewer to think, "Oh yeah, I remember that in 1986." If a designer purchases the clothes, they should still always be looked at as costumes: whether they're for stage or film, and whether they are broad or delicate, they are theatrical. Films are theatrical, no matter how wonderfully intimate they are. If they're not, where's the entertainment value?

Character development, to me, is the most interesting element of any costume design job, and it's what I like best about it. That's where the creativity comes in. It isn't from creating a dress or a suit, it's creating a new person. Many of my friends are writers, and we have a great camaraderie because we share the same interest in creating characters. Designing costumes is storytelling in the same way that a writer or a director tells a story. The right set may help the actor to create the character, but costume designers do so in a physical, intimate way. Our work goes directly to bringing forth the personality that is written on the page. As costume designers, we get under the character's skin the way an actor does.

biography

Deborah Nadoolman grew up in New York City standing in the back of the house at Broadway shows after her high-school day was finished. She was awarded the first grant for costume design from the National Endowment for the Arts. She had a baptism in wardrobe working at NBC Television Studios, where she learnt the full range of costume-related trades under the tutelage of television variety-show designers. Her

deborah nadoolman

versatility was an asset for director John Landis for his **The Kentucky Fried Movie** (1977), in which a dozen short segments were molded into a whole. More collaboration with Landis produced comedy classics **Animal House** (1978) and **The Blues Brothers** (1980). Nadoolman depicted the elaborate period of the 1940s in Steven Spielberg's **1941** (1979), starring comedy legend John Belushi. Her work with Spielberg continued with the iconic costumes for Indiana Jones in **Raiders of the Lost Ark** (1981). Nadoolman has had wonderful experiences working with French directors Louis Malle on **Crackers** (1983) and Costa-Gavras on **Mad City** (1997). Further designing for John Landis included **An American Werewolf in London** (1981), **Trading Places** (1983), **¡Three Amigos!** (1986), an Academy Award nomination for **Coming to America** (1988), **Oscar** (1991), and the groundbreaking video **Michael Jackson's Thriller** (1985), winner of MTV's first Music Video Award. She is the current president of The Costume Designers Guild, the union representing working Hollywood costume designers, and at time of press is completing her doctoral dissertation.

interview

My parents owned a camp for deaf children in upstate New York. Every week there was a theatrical night, for which we made costumes out of found objects. We could use anything— bath towels, paper placemats, sheets—every week was a new scavenger hunt. I found I was pretty good at this, even from a very early age. I met my husband, John Landis, when I was 18 years old. I had come to California for a summer visit, and we became friends. John had already made his first feature film, **Schlock**, when I arrived at UCLA in 1973 to begin my masters degree in costume design. When I graduated two years later, because of seismic changes in the business, everyone in the entertainment industry believed that costume supervisors who "shopped for costumes," would be replacing traditional costume designers on the film set.

The costume design opportunities available in Hollywood in the early 1970s were not in feature film, but in "live television." At that time, NBC had the largest wardrobe department west of the Mississippi River. It was the high point of the television variety show, such as *Sonny and Cher*,

and *The Carol Burnett Show*. The costume designer for each show received the script early in the week and had three days to create sensational beaded gowns, witty costumes for comedy sketches, foam fat suits and ingenious breakaway costumes. Angie Jones, my mentor at NBC, hired me in 1975, as a stock girl. From each costume that I returned to stock, I learned costume construction. Bob Mackie, Ret Turner, Bill Belew, Bill Hargate, and Pete Menefee were the versatile designers of the big variety shows of that era. They have never been given the credit they deserve for producing clever and beautiful costumes week after week, with very little budget, and absolutely no time, for an enormous and appreciative American home audience.

At NBC, I had a true Hollywood apprenticeship, including millinery, dyeing, beading, feathers, glove-making, shoemaking—variety shows used all of those anachronistic crafts on a weekly basis. Al Nichol and Lily Fonda at Western Costume Company, who had worked with every Hollywood star, from Marlene Dietrich to Elizabeth Taylor, and every costume designer, including Adrian, Travis Banton and Irene Sharaff, took me under their wings. I belong to a generation of costume designers who were lucky enough to work side by side with the last of the Hollywood studio wardrobe artisans, trained under the studio factory system. I had the pleasure of knowing Walter Plunkett (**Gone With the Wind**), Irene Sharaff (**The King and I**) and Dorothy Jeakins (**The Sound of Music**).

Different costumes are successful for different reasons. Piero Gherardi's sophisticated designs for Federico Fellini's **Juliet of the Spirits** are so witty. Danilo Donati's costumes for

Fellini's **Casanova** are an artistic tour de force. Everybody acknowledges those efforts as being costumes. But then, it's remarkable how Debra Winger and Shirley MacLaine's characters were revealed through their clothes, and expressed their eccentricities, in Kristi Zea's costume designs for **Terms of Endearment**. When the public is not meant to be aware of the costumes, then not being aware of the costumes is the great success. The moviegoer may discuss the film, and not be able to identify any piece of clothing, which is emblematic of the best costume picture ever made. Contemporary designers of contemporary movies suffer for their own virtuosity. They're punished with invisibility when the costumes disappear, and when the public doesn't notice the costumes. Contemporary films, like **Trading Places**, don't win Academy Awards for costume design, but the vast majority of theatrical product has a contemporary setting. I have yet to meet the designer who will go to a department store and buy the clothes that then appear unchanged on the screen. Because the costume designer has chosen, fitted, combined, changed, dyed, aged, adapted, altered, revised, and acted upon purchased clothing in a way that transforms an actor into a character, this originally purchased apparel is correctly considered a designed costume. A costume designer is creating a costume for a character. Costumes do not design themselves.

In order to create costumes for a character, a designer needs to know everything he can about that person. The more articulate a director is, the more information the designer has, and the more accurate, and more successful, the costumes will be. It's a pretty simple equation. Costume designers in the opera paint with a very big brush. Costume designers for the movies paint with a very tiny brush, always keeping in mind that each frame of film is a painting. Imagine an Edward

2

4

5

3

(1–5) **The Blues Brothers**: John Belushi and Dan Aykroyd (1) had already established their characters on the television show *Saturday Night Live* before they made the film. Nadoolman wanted to sharpen up their look and give them very defined silhouettes. She went to Dobbs' Hats in Ohio, to pick exactly the right hats, and made their shirts and suits to flatter their different body types. Queen of Soul Aretha Franklin stars in both the original film (5) and the 2000 sequel (4). Nadoolman intended the colorful costumes worn by Franklin and the three women to look like "reinvented Chanel," a strong contrast to the greasy waitress outfit she wore in the original film. Sketch and still of singer Erykah Badu as the witch Queen Mousette, wearing a costume inspired by Nadoolman's memories of Haiti and recreation of the court of Emperor Henri Christophe (2–3).

1

(1–3) **Animal House**: This campus comedy is based on the real-life memories of three writers, Chris Miller, Harold Ramis, and Douglas Kenney, who wrote about their university experiences at various American fraternities. When researching the costumes for the film, Nadoolman went back over the yearbooks, personal photographs, family albums, and home movies of all three writers to build up a true picture of characters who would populate the "Animal House." John Belushi (center) and his cohorts (1). Sketch of costumes for the toga party scene (2), and Nadoolman on the set of **Nothing But Trouble** with late comic actor John Candy dressed as a bride (3).

2

3

Hopper painting: Hopper is the director; the background is the production design; the cinematographer supplies the lighting; and the costume designer designs the costumes for the people who live in that world, within that proscenium. The better the communication is among the director, production designer, cinematographer and costume designer, the more cohesive the look (and invented world) of a film. Each frame of film represents one canvas. It's not casual. Every frame of film is minutely handcrafted.

When I left my interview for **1941**, Steven Spielberg announced to me that it would be a "Nadoolman-Spielberg co-production." Although it was a commercial failure, it was a delight to design and a pleasure to work on. Spielberg's door was always open, and he was always receptive to my ideas. When we started **Raiders of the Lost Ark**, I met with Steven and Larry Kasdan, who wrote the script. Steven drew a charming, childlike sketch of Indiana Jones for me, 6 feet, 2 inches with a little hat, jacket, and a bandolier of bullets, and was very clear about the character. **Raiders** was a remake of **The Secret of the Incas**. He said it would be a B movie with an A budget. I reinvented, cleaned up, and codified Charlton Heston's **Incas'** costume, including his well-worn brown leather flight jacket—to which I added a useful "action back," and khakis for Indiana Jones. I designed Indiana's classic brown felt fedora for Harrison Ford, adapted from a model I found at the very British London hatter Herbert Johnson. I lowered the crown and reduced the width of the brim to let more light onto Ford's expressive face and eyes. Then, using Harrison Ford's Swiss Army knife, a steel brush, and many sheets of sandpaper, I destroyed my hands by aging Indiana's first leather jacket myself (of the ten jackets we manufactured), poolside at La Rochelle, the day

before our first day of shooting at the U Boat dock. When Ford arrived on the set, it looked as if he had been wearing that jacket for 20 years. Any costume that can be instantly recognized in silhouette has transformed into an icon. The great success of the costume of Indiana Jones personifies the perfect communication between an actor, a director, and a costume designer.

In **Coming to America**, Eddie Murphy portrayed Prince Akeem as an Oxbridge-educated African prince. Though we were making a comedy, his clothes were seriously elegant. They weren't meant to be a sight gag, like the costumes in **The Three Amigos**, but to highlight Akeem's Western education and African roots, they brought a tremendous amount of information to Eddie Murphy's characterization.

Approaching **Coming to America**, I knew I would be using hundreds of yards of African fabric, and there was none to be found in the US. Every fabric was ordered from England, through a market stand in Brixton, that I had discovered while designing **An American Werewolf in London**. Richard McDonald, the production designer, and I established a style for the kingdom of Zamunda, before we started manufacturing the sets, and costumes. The design of Zamunda was based on the fantasist early 19th-century palace, the Royal Pavilion in Brighton in the UK, and the naïve *Peaceable Kingdom* paintings of Henri Rousseau. It was the richest kingdom in the world, and, like Brigadoon, appeared from the fog behind the Paramount mountain. Everything about Zamunda was beneficent; James Earl Jones was the enlightened monarch who reigned over an adoring population. We contrasted life at the palace with the world of Queens, New York, and there would have to be a balance. Queens was not all ghetto and

costume design

(1–7) **Coming to America**: Directed by John Landis, this delightful comedy stars Eddie Murphy as His Royal Highness Akeem, Crown Prince of Zamunda, who reacts to the news that his father is arranging a marriage for him by coming to Queens, New York, to find his own bride. It is implicit in the background of the film that Prince Akeem was educated in England, and Nadoolman used the tailoring of his clothes—shirts and ties made at Turnbull and Asser in London's Jermyn Street—to provide clues to the character. Sketch and still of Zamunda dancers (1, 3), and sketch and still of Prince Akeem (2, 4). For the art direction, production designer Richard McDonald and Nadoolman were inspired by Fauvist art, the paintings of French artist Henri Rousseau and the spirit and architecture of the Royal Brighton Pavilion, UK (6).

5

7

6

poverty, although there was plenty of opportunity to take advantage of the comedic clash of cultures. **Coming to America** is the African-American Cinderella. It was a privilege to have had the opportunity to work with John Landis on his groundbreaking, color-blind film.

My assistant designer and sketch artist, Kelly Kimball, and I have worked together since 1975. We share a passion for research, ethnic costume and textiles. For **Thriller**, I designed Michael Jackson's costumes, and Kelly worked on the ghouls. **Thriller** set the stage for the elaborate music videos that followed, and its influence in costume and production design still resonates. Michael Jackson is one artist who understands the power of a costume.

The human eye sees in three dimensions, but film is two-dimensional. A costume designer must look at costumes as a camera would. Often, I will look at fabric reflected in a mirror, because that's how a camera sees it. The costume designer compensates for how fabric is diminished on screen. An old mantra of costume designers is, "Increase the scale by 30 percent," because you lose the third dimension.

Vintage clothes may not "read" brightly enough on screen; they may need to be "pushed." An authentic lace dress from 1905, beautiful in your hand, has detail so fine, it will never be seen on screen. A designer will augment it in some way, perhaps by exaggerating the silhouette of the sleeves, increasing the scale of the lace, or creating additional depth or shadow. Real clothes can often look drab, and flat, on screen.

The first feature film I designed was **The Kentucky Fried Movie**, a wacky, low-budget comedy with a costume budget of

$15,000. Then I designed **Animal House**, with a budget of $50,000, then **1941**, with a costume budget of $250,000. Regardless of the budget, the job of costume design was exactly the same, and the chronology of the work was identical on each movie. **Animal House** was based principally on writer Chris Miller's early 1960s Dartmouth fraternity. The director, John Landis, wanted the "Animal House" to be distinct; a good, inclusive (although slovenly) fraternity, and the WASP Delta fraternity, the "bad guys," to look extremely uptight and preppy. I used a strict color palette to separate these two groups. The fraternity toga party existed, but the "Animal House" togas (and John Belushi's laurel wreath), were pure invention, painted by Judy, John's wife, and myself. My early experience at camp, making costumes from sheets, came full circle. Although **Animal House** was low budget (I personally sprayed all of the rubber flip-flops gold), I am confident that the costumes were as carefully designed as any serious period film made at the time. There was a scrupulous attention to detail. Two of the writers appeared in the movie. We had recreated the fraternity of their youth. They relived it in my costumes, which was a great compliment to my work.

John Belushi and Dan Aykroyd had established the Blues Brothers characters on the television show *Saturday Night Live*. They had thrown together a costume that was emblematic of blues singers in the 1950s, based on the work established by John Lee Hooker. For the film, I sharpened that look, by creating a defined, iconic silhouette. I looked for the right fedora at Dobbs' Hats, when Dobbs' still had a factory in Ohio. I made their threadbare white shirts, their hats, ties and suit coats, which are not identical. Creating styles that flatter the actor are another hidden element of

1

2

3

4

(1–4) **Oscar**: Set in the 1930s, and intended as a homage to the great screwball American comedy films of that era, **Oscar** is the tenth film in Nadoolman's long collaboration with director John Landis. The film stars Sylvester Stallone as Angelo "Snaps" Provolone, a gangster who promises his father on his deathbed that he will go straight—a promise that incurs hilarious consequences.

1

(1–4) **Raiders of the Lost Ark**: Nadoolman feels that she had a very productive relationship with director Steven Spielberg when designing costumes for this film. "I felt his door was always open to me, which is extremely helpful for a costume designer. He made himself available for questions, and so I didn't feel like I would ever surprise him unfavorably with a costume on a set." Nadoolman's costume for Harrison Ford as Indiana Jones has become an enduring and iconic image of cinema.

2

3

4

costume design. Jake and Elwood Blues were meant to look sleazy, and also attractive and sympathetic. **The Blues Brothers** was a classic Hollywood musical, not a documentary, and the costumes belonged to that genre. I needed as many pairs of Ray-Ban Wayfarer glasses as I could find, with so many car stunts, and stunt doubles. When I called Ray-Ban, it had stopped manufacturing them. The style had remained popular in the African-American community and the only place to find them was in Watts and in Harlem. I found 25 pairs of Wayfarers, but every time John Belushi flirted with a girl, he took off his glasses, and gave them to her. Though he was a delight, I was constantly looking for replacement glasses.

When a costume designer is permitted, he can contribute a tremendous amount to the creation of the characters in a movie. In **Blues Brothers 2000**, Aretha Franklin played the owner of a Mercedes Benz dealership. After appearing in a stained pink waitress uniform in **The Blues Brothers**, I imagined her and her backup singers in head-to-toe wild, exuberant, Chanel-inspired suits, with chain belts, quilted handbags, and cap-toed shoes. Aretha loved it. The costumes for Erykah Badu, playing a Cajun witch, Queen Mousette, were based on my childhood memories of Haiti and King Henri Christophe. I had an image of a Marie Antoinette wig that was made of dreadlocks, and an 18th-century dress of African fabric. Her bodyguards were inspired by recollections of seeing "Papa Doc" Duvalier's thugs, the Tontons Macoutes, in tight new jeans, work shirts, straw cowboy hats and boots, and aviator sunglasses. They were absolutely terrifying. Although John Landis might have expected this "chorus" to be dressed in generic fatigues, he was delighted at how the costumes added to the real threat of Mousette's private army.

Given enough freedom, security and artistic partnership, a costume designer can help create the look of a movie.

Visiting a Paris fabric show with a close friend who is a 7th Avenue designer, I was shocked when she bought 10,000 yards of corduroy. She said, "But this is so similar to what you do." I countered, "No! I am always looking for the perfect three yards—the three yards that will look right for the character, and the camera." I am creating a costume for one moment in the arc of a story. Costumes are created for a single purpose, for one use only.

Fashion and costume have different, antithetical purposes. Fashion designers sell clothes. Costume designers create costumes to sell a character; to support actors in their craft; to help convey a story. Costumes cannot be judged outside of the context of the scene for which they were created. They have nothing to do with real life, so when a costume is put on exhibit in a museum gallery, it can't be judged in the same way as the Versace dress next to it. A Versace dress is meant to be worn by a real person, seen in three dimensions, the way you're seeing it in the museum. The costume was created to appear under certain lighting, through the eye of a lens. It was meant to be seen big. It was made to be worn by an actor, in the context of a set, within a story that you do not experience in the museum. What the viewer sees in an exhibited costume is an empty shell, with no soul or purpose. Costume designers don't think about whether their costumes are going to generate a million copies after the film comes out. They ask themselves, "How can I tell the truth with the costume that I am creating for the character?"

Gabriella Pescucci has worked with Italy's most highly respected filmmakers in the creation of movies noted for their visual style. Early in her career, while assisting Piero Tosi and Umberto Tirelli, she was given the opportunity to design costumes for the films of the Italian master filmmakers Luchino Visconti, Mauro Bolognini, and Federico Fellini. In the 1980s, her career was marked by repeated collaborations

gabriella pescucci

with Ettore Scola, with whom she worked on **Passion of Love** (**Passione d'Amore**, 1981), **That Night in Varennes (La Nuit de Varennes**, 1982), **The Family (La Famiglia**, 1987), **Splendor** (1988), and **What Time is It? (Che Ora è?**, 1989). Pescucci earned her first Academy Award nomination for her brilliantly theatrical costumes in Terry Gilliam's fairytale **The Adventures of Baron Munchausen** (1988), and won an Oscar for a serious period portrait of upper class society in turn-of-the-century New York, for Martin Scorsese's adaptation of Edith Wharton's book *The Age of Innocence* (1993). She also won a BAFTA award for Best Costumes for **Baron Munchausen** and for her marvelous contribution to Sergio Leone's classic tapestry of American gangsters, **Once Upon a Time in America** (1984). While the majority of Pescucci's work has been for the big screen, she has designed costumes for several of Liliana Cavani's beautifully executed operatic masterpieces for television, including *La Traviata* (1992), *Cavalleria Rusticana* (1996), and *Manon Lescaut* (1999).

interview

I was very fortunate to have begun my career with such exceptional artists. I had been interested in becoming a costume designer since I was very young. After finishing my studies in Florence at the Academy of Fine Arts at the end of the 1960s, I went to Rome in order to pursue a costume designing career. I met Umberto Tirelli, who had opened his famous costume house a few years earlier. He hired me to work in his shop. All of the designers working at the time would come occasionally to Umberto's, and I remember being very excited when I saw that Piero Tosi had come in. I worked with Umberto for a few years and I learned all sorts of things that it would have been impossible to learn at school. Eventually, I mustered up the courage to approach Tosi on one of his visits, and he took me on as his assistant.

Being mentored by Piero was truly special. From Piero and Umberto, I learned everything I know about designing costumes. They were both great teachers (and great friends). Moreover, as Piero's assistant, I had the opportunity to work with the great directors of the Italian cinema. With Piero,

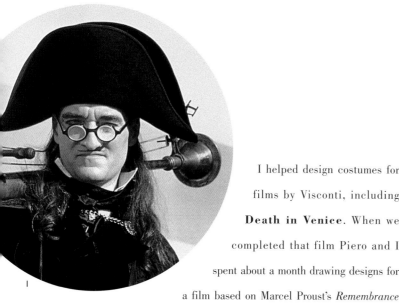

I helped design costumes for films by Visconti, including **Death in Venice**. When we completed that film Piero and I spent about a month drawing designs for a film based on Marcel Proust's *Remembrance of Things Past*. We never read Visconti's screenplay for the film; the script was not complete, and he gave us "pieces" of it from which to work. Eventually the project was given up. It just seemed too difficult to make a two-hour film out of such a long, complicated story.

I recently designed the costumes for Raoul Ruiz's version of the story, **Le Temps Retrouvé**. Raoul had the courage to include just a part of the book. He felt free to play with the narrative—even in the face of the French critics, who couldn't wait to lambaste his efforts. Raoul is a great intellectual. Working on this film was particularly interesting because of the South American fantasy element of the design. The style bore a resemblance to that of Buñuel.

Piero taught me was that it is all right to make mistakes and to learn from them. More importantly, he taught me that dramatic effect takes precedence over historical accuracy. After **Death in Venice**, I assisted him on Visconti's **Ludwig**. I was working on some uniforms for the film and he said to me, "No, Gabriella, I researched these uniforms, and they were a really awful blue. I'll *never* use that blue. Let's change it, let's make them a more believable blue." I was awestruck by him. Piero knew how to bend the truth—and rarely does anyone notice. I look back on the costumes I helped Piero create for Pasolini's **Medea**. They were the result of countless different sources and inspirations.

Independently, I designed the costumes for two Fellini movies, **Orchestra Rehearsal** and **City of Women**. Being able to work with Fellini was a great joy for me. I felt very fortunate. To describe what he was looking for, Fellini wouldn't go into detail, he would simply use concepts. When we were making **City of Women**, his image for the feminists was a fruit jam. I remember it very well. He meant that he wanted to see a mass of women moving. He wouldn't say, "I want the women dressed in red," or, "Put the feminists in green." He would wait to see what the designer created.

One of my best memories as a professional is working with Sergio Leone on **Once Upon a Time in America**. Leone had a frightening presence: he was very big, very physical, and tough. Before I accepted the job on the film I told him: "Mr Leone, they tell me that you yell a lot on the set. Please know that I am very sensitive and if you yell at me once, you'll never see me again!" He laughed, and we had a beautiful relationship. Underneath the surface, he has a rare combination of artistry and common sense. He had a wonderful understanding of nuance; he could understand textures. I would say to him, "Sergio, come and feel how soft this silk is," and he would understand what the softness of the fabric meant. It was most extraordinary in someone who was so rough and so practical.

My work depends completely on the movie and what the director asks for. For **Once Upon a Time in America**, Sergio wanted verisimilitude, and I think that his America of the 1920s was very believable. Some critics have said that the film had a look that is too European, which may be true—after all, everything is filtered through one's culture and sensibility. Sergio had collected a huge archive of beautiful photographs of

costume design

2

3

4

(1–4) **The Adventures of Baron Munchausen**: A spectacular fantasy extravaganza directed by Terry Gilliam, one of the members of legendary British comedy group Monty Python. It tells the story of the fictional 17th-century European aristocrat Baron Munchausen and his adventures. This was the fifth film on which Pescucci worked with production designer Dante Ferretti. Their winning partnership won both of them a BAFTA and Academy Award nominations. Pescucci with cast in costume (4).

1

2

3

4

5

(1-9) **The Age of Innocence**: Martin Scorsese's masterful examination of polite society and suppressed eroticism in 1870s New York. Winona Ryder is May Welland, a socialite poised to marry lawyer Newland Archer, played by Daniel Day-Lewis (1). May's unconventional cousin Countess Olenska, played by Michelle Pfeiffer, upsets the balance when Archer begins to fall in love with her. The costumes perfectly express the repressive, etiquette-obsessed life, which the characters are reined in by. Sketch, Pescucci with dress on model, and production still of one of May's dresses (3–5). Stills and sketch of Countess Olenska, whose costumes symbolize her outsider status (6–9).

7

8

6

9

gabriella pescucci

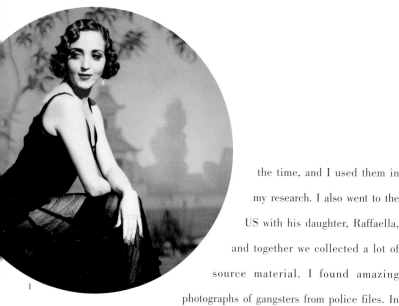

the time, and I used them in my research. I also went to the US with his daughter, Raffaella, and together we collected a lot of source material. I found amazing photographs of gangsters from police files. In addition to real stills, we relied on the style of films from the 1920s and '30s. Sergio intended the film to serve as an homage to those movies and their heroes. He was maniacal about the Borsalino hats that the gangsters wore. We had one old worker from Borsalino's on the set, and he was constantly pressing the hats. They had to be perfect. The American gangsters were very elegant, and Sergio insisted that the tips of their hats be perfectly rigid. Even in difficult moments, there was a great harmony among the crew on this film. I have happy memories of working on it and of Sergio's kindness.

I made quite a few movies with Ettore Scola. I loved working with him on his masterpiece, **The Family**. So much of this movie was created through improvisation—every two weeks or so, we would start shooting another historical period. Ettore would point out a certain character and ask, "Shall we make this one die? No? Okay, let's have her live another decade!" Then we would age her, we would add some wrinkles, or have her posture become more stooped. We collaborated very closely on this film, and it was a lot of fun to be on the set.

It was about this time when I began to feel a little bit lost as a professional. The great directors of my youth were no longer making films. During the 1970s and the early '80s, I felt no desire to work outside Italy, but over time I realized that I would have to find new directors to work for. I was lucky to meet directors from all over the world. In the late 1980s, Terry Gilliam came to Italy to film **The Adventures of Baron Munchausen**. I was fortunate to be hired to design the costumes. This was the first international type of movie that I worked on, and I look back on it as one of my favorite projects. Working with Terry was a great experience; I had a sensation of the utmost creative freedom. Terry is such a curious person, and he is very open-minded. If I offered an idea that was different from what he asked for, he would always consider it. Often he would implement my suggestions. He's very knowledgeable about culture—painting, sculpture, architecture—and he uses this knowledge in creating a style for his films. This was very helpful to me in doing my job as the costume designer. I received my first Academy Award nomination for **Baron Munchausen**, which raised my profile professionally.

After **Baron Munchausen**, Terry began working on a film version of *Don Quixote*. I had completed about 90 percent of my designs for the movie when the project was discontinued. It was disappointing, because the film was being made very much in Terry's visual style, with modern elements being intertwined with Don Quixote's world. We worked the way we had before: we started with the historical period and then he said to me, "You take it from here, feel free." As a costume designer, this is the most liberating type of direction. With instructions like this, you get to create a new genre; you get to make intentional mistakes. I love making mistakes and ignoring dates—it is a sign of great artistic freedom. Over time, I appreciate this kind of stimulation all the more, and am happy to make mistakes whenever I'm asked to.

Finding the right fabrics for costumes, particularly period pictures, is becoming increasingly difficult. Sometimes it

(1–4) **Indochine**: Set in 1930s Indochina, the film is a sweeping historical romance based in the south-east Asian peninsula. Catherine Deneuve is brilliantly cast as the French rubber plantation owner Eliane Devries who falls in love with a naval officer (2–4), only to have him begin a long liaison with her adopted Vietnamese daughter. Pescucci shares credits with costume designer Pierre-Yves Gayraud on the elegant costumes, which blend beautifully with the stunning Vietnamese landscape.

costume design

(1–4) **A Midsummer Night's Dream**: Michael Hoffman's 1999 adaptation of William Shakespeare's festive comedy with a winding plot. Sketch for fairy queen Titania and members of her kingdom (1). Michelle Pfeiffer as Titania in one fantasy costume (2), and Pescucci with the same dress on a model (3). Kevin Kline as Nick Bottom, the simpleton who is transformed into a man with a donkey's head (4).

seems that the only appropriate fabric I can find is for upholstery, but then it is often so heavy that the costumes look like walking armchairs. You can only find materials that are in today's style in the fabric shops—and with fewer and fewer people making their own clothes, I'm afraid that the shops themselves will just disappear.

We had to do a great deal of aging to create the right look for the costumes in **The Name of the Rose**. The friars were dressed in black, but not a true black: it was a brownish-green fabric that we tinted black and then re-bleached. I stayed in Rome, and Tirelli left his shop open for me and my assistants to work in. We tinted a lot of heavy woolen cloth that absorbed volumes of color. When he came back and learned how much I had spent on colors and tints, something like $7,000, it was a scandal! **Les Misérables** was also an enormous tinting and aging job. We patched so many costumes for that film, you'd think that they had been inspired by Burri (the Italian Abstract Expressionist painter).

Sometimes it's harder to do a good job designing costumes for poor people than for wealthy people. For rich characters, once you find the fabric, you can get to work. For poor characters, you start with the cloth, then you tint it, then you sew the costume, then you tint it again. It's harder to get it right, and it's more time-consuming. I like de-colorizing the costumes and experimenting. It's like painting or drawing. Regardless of the characters I'm dressing, my greatest satisfaction comes from having my work disappear in the film. If it doesn't jump out at the audience, it's serving the purpose of the movie.

I like to approach designing costumes for a period movie with an open mind, and allow myself to take in every avenue to the era.

To research how to best represent a period through costumes, I'll look at paintings from that time. They're an endless source of inspiration. I'll examine old photographs and watch old movies, and spend a good deal of time looking at books of art and sculpture. Sometimes I research the architecture of a time period for a sense of the shapes that were popular. Novels written during the movie's period are often full of suggestions. The prevalent lifestyles and behaviors are very informative about what people wore. When I feel that I've done a fair amount of research, I start to draw. Drawing helps to plant images in my mind so I can elaborate on them later. After that, I start searching for fabrics.

Several aspects of costume designing have changed over the years. One big difference is the amount of time you get to prepare for a film. The time between when a picture is green-lit and the time you actually start shooting seems to get shorter and shorter. Movies have gotten so expensive to make, it could be that no-one wants to spend any money until they're certain that all the pieces for the production are in place. Another change is simply the size of people. Not that long ago you could rely on old costumes. Now, costumes that are 20, or even 10 years old don't fit the extras. You can't find a woman who wears a size 38 shoe. Of course, the changes in technology affect everyone in the moviemaking business. One thing that doesn't seem to change is that you can't learn this craft in school. The best way to learn to design costumes is by doing it. The need to be in love with your profession is timeless. You must find joy in seeing a character come to life from your work.

biography

The consummate collaborator, Anthony Powell has said that costume design demands "a sympathetic understanding of others' points of view in order to achieve a cohesive whole, whilst still being able to make a positive statement of one's own. A talent for designing is, in itself, only 50 percent of the job." Powell has designed costumes for over 20 feature films and an equal number of stage productions. He won his first

anthony powell

interview

Academy Award for one of his earliest movies, **Travels with My Aunt** (George Cukor, 1972). Powell then dressed Steve McQueen and Dustin Hoffman in Franklin J. Schaffner's gritty prison drama **Papillon** (1973), followed by Robert Altman's Western extravaganza **Buffalo Bill and the Indians** (1976), starring a blonde, buckskinned Paul Newman. Shortly thereafter, Powell garnered two more Academy Awards for perfectly realized period films: **Death on the Nile** (John Guillermin, 1978) and **Tess** (Roman Polanski, 1979). **Tess** served as the foundation of a Powell/Polanski partnership, and the two have since worked together on **Pirates** (1986), for which Powell was nominated for an Oscar, **Frantic** (1988), and **The Ninth Gate** (1999). Other credits include three movies with Steven Spielberg: **Indiana Jones and the Temple of Doom** (1984), **Indiana Jones and the Last Crusade** (1989), and **Hook** (1991). Powell's work has been widely admired by other designers, and his generosity of spirit treasured by many. He was named Britain's Royal Designer for Industry, and honored by the Costume Designers Guild with its Career Achievement Award 2000.

Having spent my entire childhood obsessively making model theaters and putting on shows, by my late teens I was preparing (without enthusiasm) to take up a scholarship in modern languages at Trinity College, Dublin. Then, in a life-changing moment, my stepsister met a wonderful man named Christopher West, who was resident director at London's Covent Garden Opera House. He suggested that I send my portfolio to him. He was very encouraging, and arranged my admission to the Central School of Arts and Crafts, probably the finest place in the world to study theatrical set and costume design.

I did their three-year course, studying with Norah Waugh, the pioneer researcher into the evolution of the cut and construction of period clothes. I was taught how to *look and see*, and given a sense of form, volume, spatial relationships, and proportion, through the drawing, painting, sculpture, and architecture classes—all neglected in present-day English art schools, unless one is actually studying fine art. Strong influences were the work of Christian Bérard, Antoni Clavé (with his stunning ballets for Roland Petit), Oliver Messel and Cecil Beaton (both of whom later employed me as an assistant), and Lila de Nobili, who demonstrated to designers of my generation how to create transcendentally magical poetic realism.

Throughout the 1960s I designed sets and costumes for operas and plays, and had consultancy jobs in industry. One evening I met the film director Irving Lerner at a dinner party. The next day he asked me to go with him to Spain to design the costumes for an epic about the *conquistadors* in 16th-century Peru, called **The Royal Hunt of the Sun**. When students ask me how to break into films all I can say is, "Go to lots of dinner parties."

By making every mistake in the book, I was taught 90 percent of the lessons to be learned. One of the most important was never to listen when you're told, "Oh don't worry—nobody's ever going to see it." The furthest extra in the background needs to be given as much thought and attention as the star, as, likely as not, the camera will end up on that person. You have to do much more work than will ever be seen on screen to cover yourself. Particularly in period films, if extras have the right faces you need to do very little to make them look authentic, but if not, you can throw fortunes of money and time at them to no avail. This is why I like to choose the crowd artistes myself.

While making **The Royal Hunt of the Sun**, I naïvely believed it when told that there wasn't anyone in Spain who made hats or jewelry, so everything in the film was made by me! We were based in Madrid, and due to shoot a scene at daybreak on top of a mountain in Granada. The sequence involved many Incan priests covered in crowns and jewelry that hadn't yet been made. I sat in the car with an anvil, hammer, and sheets of tin, beating them out as we drove through the night. When we got to the bottom of the mountain I could see everyone at the top, so I told the driver to drive terribly slowly up the winding road, and we arrived with the final crown just finished.

Another important lesson I learned early in my career was not to accept scripts simply because they gave you a chance to design spectacular costumes. The overriding factor in choosing a project is the director. A good director will have a vision, and the idea of the movie will come from him. It can't come from someone on the perimeter, such as the production designer, or the cinematographer, or the costume designer. It may be flattering to be given such authority, but it never works. It's all got to come from the center: it's like a wheel going round and everything must radiate from the hub.

George Cukor was the best director I've ever worked with in terms of his ability to describe the atmosphere of a scene, or the way he wanted a character to look. He'd never say "I think she should be in pink," but you'd come out of the meeting knowing exactly what he wanted to see. Knowing which path to follow, I still had total freedom. He was a genuine Hollywood legend and I was a beginner, but he treated me as an equal. I was helping prepare Maggie Smith to go on set when he called to ask, "Which side of the staircase would you like me to put the camera?" I was so startled that I stammered, "On the right, please." All the detail of Maggie's costume was on the left side, so on screen you saw nothing. He would re-block and re-light a scene to accommodate an unexpectedly striking outfit.

1

2

3

(1–3) **Buffalo Bill and the Indians**: Based on the real-life showman William Frederick Cody (1846–1917) who toured Europe and the US with his Wild West Show, this stars Paul Newman as William Cody, aka Buffalo Bill. When researching the costumes, Powell found a tattered old buckskin jacket at the back of a cupboard, bead-embroidered with American flags and flowers, felt it would be perfect for Cody's show costumes, and had it copied for the film. Many years later, Powell discovered that the original had in fact been made for one of the real Buffalo Bill's appearances at Madison Square Garden, New York in the 1880s. The Stetson hat worn by Paul Newman is also a direct copy of the original hat Buffalo Bill used to wear in advertisements (1–2).

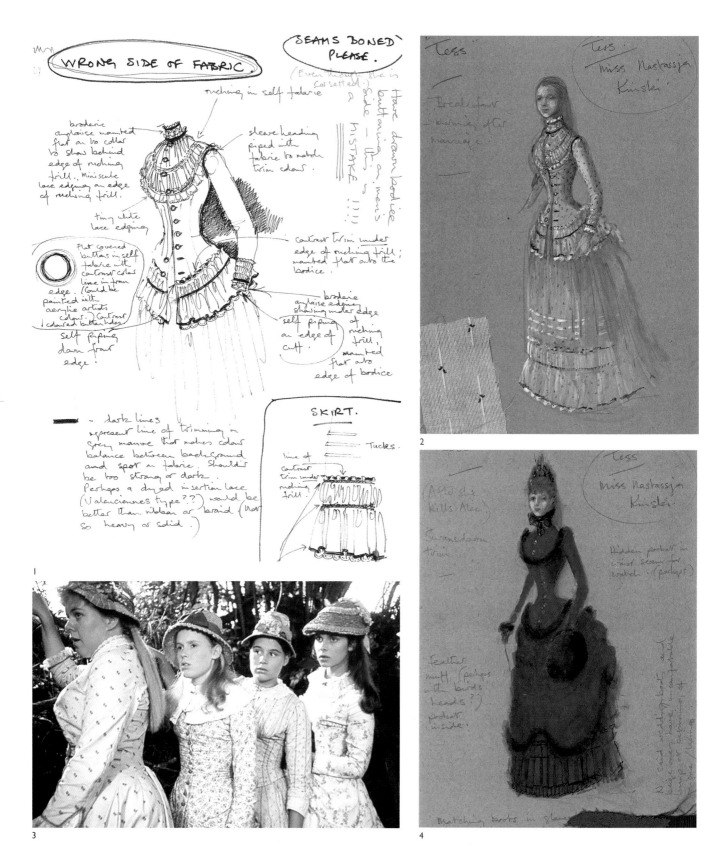

(1–5) **Tess**: Roman Polanski's adaptation of Thomas Hardy's classic 19th-century novel *Tess of the D'Urbervilles*. Rather than shooting the film in Dorset, England, where the novel is set, the whole film was shot in Normandy and Brittany in France. The small, domestic-sized fields and real-life peasant workers were more redolent of 19th-century rural England than anything Polanski could find in the UK. "I used only the palest colours (which registered as off-whites) and earth tones, which disappeared into the landscape (3), until the very last moment, when Tess has murdered her protector Alec d'Urberville, and she suddenly appears in a traveling outfit the color of congealing blood (4–5)." Powell designed the costume around a short length of original Victorian silk plush, chosen for its depth of color. (1–2) Detailed notes and sketch of one of Tess's day costumes.

Designing costumes is most fulfilling when a director is flexible. For example, when Steven Spielberg comes to the first production meeting of a film, you feel that he's already shot the film in his head. But I found him very open, and if you have an idea that he likes, he'll work it into the shooting. Originally, **Indiana Jones and the Temple of Doom** was supposed to open with the titles running over Kate Capshaw singing in a Shanghai nightclub. I suggested expanding this sequence into a sort of Busby Berkeley number. It would be fun for Kate, and it wouldn't interfere with the narrative. I did some sketches and he said, "Yes, great, let's do it," and we did.

Robert Altman is infinitely adaptable. On **Buffalo Bill and the Indians**, if an actor spilled coffee down the front of his shirt, he'd say, "Fine, that's real life—you'll play the rest of the film with coffee stains on your shirt." Geraldine Chaplin, who was playing sharpshooter Annie Oakley, broke her right arm the day before we started shooting, so she played the entire film with her arm in a sling, and did all the rifle-work with her left hand.

In contrast was David Lean, with whom I worked for about two years on his last project, Joseph Conrad's *Nostromo*. Sadly, Lean died before it could be made. He was a master of cinema; it flowed from his fingertips. His production designer, John Box, had been a mentor throughout my career so I came to David already "housetrained." Once he had completed a script, the entire film, with every camera angle, existed in his mind, and that was all he wanted or would accept. At that stage, no one else could contribute. He once said to me—but not *of* me—"I can't bear it when people start getting *creative*!" I was lucky enough to be involved from day one and had

found extraordinary research material that he loved and incorporated in the script.

One of the happiest projects was Roman Polanski's **Tess**. We shot for a year with every kind of problem: it was the middle of winter, muddy and cold, and the money ran out, yet it was the most wonderful experience because everyone believed passionately in the project and we all worked together. For a designer, it was a dream job. Roman knows everything about all the departments, but claims that he doesn't understand costume. This isn't true, because he immediately knows whether an idea is right, and he films any interesting detail, even if it doesn't make it into the final cut. Although we have little in common, in work we have an extraordinary telepathy. We hardly need to talk. He'll phone me with a new project, and with just one sentence will give me all I need to know about his vision. For **Pirates**, he said, "I love all those huge wigs and hats," and that was it. I knew exactly what was in his mind. We've collaborated on seven projects (two of which were cancelled). The most recent was **The Ninth Gate**, where I had the real pleasure of working with Johnny Depp.

When I switched from designing theater to designing film, I realized that you can impose a character on an actor in the theater: with the distance between the actor and the audience, he can be convincing. You can take a young, thin actor and you can pad him and give him the right makeup, and he can play Shakespeare's Falstaff. But it always seems to me that in cinema, you can't impose characters on actors. You have to use what's there, because the camera sees untruth. You have to take the inherent qualities of an actor, and the character demanded by the script, and try to bring the two together somewhere in the middle. I always try to meet an actor for the first time over a meal, so that I can get a sense of who he or she really is. It is useful to see his or her own clothes, too—I get ideas for styles or colors, even for period films.

If you think of the great stars of the old studio system, many of them were stars because they had this mysterious, indefinable quality that the camera loved and responded to. Whether her character was nice or nasty, Joan Crawford was always Joan Crawford. That was what you paid to go and see. Nowadays, such actors can be very difficult for designers to work with, because they are selling a product—themselves—and one can't do anything that cuts across what they're selling. They get miscast, usually because executives nowadays rarely create projects specifically tailored to actors' qualities. The costume designer then has a big problem, because he must try to turn these stars into something that they're not. Although initially they will sometimes be willing to work with you, just before filming they usually lose their nerve and will want to go back to the way the public knows them. In the end, you wind up with a compromise that no-one is happy with.

The other type of actor, like Dustin Hoffman, for instance, or Glenn Close, or Johnny Depp, are stars because they are wonderful actors and they love transforming themselves. They are rewarding to work with, because together you're creating a character. I've loved working with Dustin, because he'll do anything that's needed for the part. For one prison uniform in **Papillon**, he stood in a fitting room for five hours in order to achieve the effect of diminishing his own robust physique. His character was described as wearing these huge spectacles with enormously magnifying lenses that made his eyes swim.

I thought this could be a wonderfully graphic image, but was concerned that Dustin wouldn't be able to see, so I suggested that he wear contact lenses that were the opposite of the prescription of the glasses. He'd never worn contacts, but persisted, and it worked. It's that willingness to try anything that is so inspirational.

That's also true of Glenn Close, which is why I love designing for her. She's as fearless as a lion. We'd worked together on the Broadway musical of *Sunset Boulevard*, and she asked me to design Cruella De Vil's clothes for **101 Dalmatians**. When I asked for her thoughts on the character, she said, "You just design it, and at the end I shall look at myself in the mirror and then I shall decide how to play the part." I thought, "Oh my God!" It's a tremendous responsibility to persuade an actor into a very extreme look. If they trust you and you get it wrong, they will actually give the wrong performance.

The first time I met Bette Davis, I had gone to her home in order to discuss her role in **Death on the Nile**. "First of all," she said, "you need to see what the problems are," and although she was 70 at the time, she whipped off her dress over her head, and there she was, virtually stark naked in the middle of the floor. Bette told me that when an actress signed a long-term contract with a Hollywood studio in the old days, she was dressed in something revealing, like a swimming costume, and was then asked to stand on a revolving plinth in the middle of a sound stage. The platform was lit ruthlessly with floodlights, and as it revolved, all of the department heads, the cameramen, the makeup artists, the hairdressers, the designers, would sit around it with clipboards, making notes about every single defect—the bust, the hips, the nose.

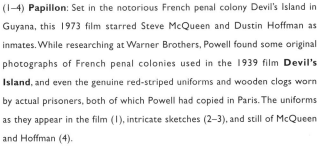

1

2   3

(1–4) **Papillon**: Set in the notorious French penal colony Devil's Island in Guyana, this 1973 film starred Steve McQueen and Dustin Hoffman as inmates. While researching at Warner Brothers, Powell found some original photographs of French penal colonies used in the 1939 film **Devil's Island**, and even the genuine red-striped uniforms and wooden clogs worn by actual prisoners, both of which Powell had copied in Paris. The uniforms as they appear in the film (1), intricate sketches (2–3), and still of McQueen and Hoffman (4).

4

1

3

(1–7) **Death on the Nile**: During preparation for the film, it was impossible to get all the cast in one place for fittings, so Powell insisted on taking two trips to see each actor, who were as far flung as India and Tokyo. Stills and sketches of Peter Ustinov and Maggie Smith's costumes (1–4). Mia Farrow as Jacqueline de Bellefort (7), and Bette Davis as the cold Marie Von Schuyler (5–6). "One of the first things Bette said to me when we first met was, 'I've spent all my life in films wearing hats, so I don't want to wear any in this one.'" However, Powell's designs had her almost permanently in hats, so he had to persuade her into thinking otherwise. A photograph signed by Davis, expressing her thanks to Powell for his designs (6).

2

4

5

Tony dear,
I will never
be able to perform
again without
one of your beautiful
huge black hats.
love,
Bette

6

7

1

(1–4) **101 Dalmatians**: "Glenn Close was a real trouper, as it involved all the most difficult and uncomfortable clothes, like corsets and five-inch spike heels. She always said, 'I don't know if I can manage it, but I'll try.' And in every case, she made it work." It was impossible for Close to sit down on set while in costume, so Powell had some old-fashioned slant boards made for her. Her character as Cruella de Vil was designed to be as extreme as possible: "It worried me to see pink hands, which seemed too human and vulnerable, so we decided she would always wear gloves (even in bed), which extended her silhouette and accentuated each gesture." Still and sketch of black and white costumes (1–2), and stills of two other outlandish Cruella de Vil outfits (3–4).

2   3   4

She said it was the most appalling, humiliating experience, but it paid off in the end because one knew that one would never be seen at a disadvantage on the screen. Everyone at the studio was invested in preventing the public from seeing any type of physical imperfection one might have. Apparently Bette had become accustomed to this type of scrutiny, and she was absolutely right to have disrobed, because then I could design costumes that would celebrate the good points and draw a discreet veil over whatever small flaws might exist elsewhere.

I am fascinated by people and feel there's virtually no-one you can't get on with if you really try—a prerequisite to this job. As a student, working with Oliver Messel, I observed that he had eliminated the word "no" from his vocabulary. Say, "No," and a steel shutter descends. There was always a way of getting what he wanted while making artistes feel that they were getting what they wanted too. It's thrilling to build characters with actors; to invent their histories and psychological makeups, so that a story is being told. If a great couturier designs the clothes for a film, then the performers are simply actors in beautiful clothes. There is no character, unless the *essence* of the person is that they are rich and wear Christian Dior ballgowns as part of their life. Then it may work.

Another aspect I enjoy, together with doing the research, is the part I think of as visually orchestrating a script, and of finding, as it were, the right musical key for the production. Then one uses color and shape, and, most importantly, tonal values, to give it highs and lows and plateaus and climaxes. It's very satisfying, like composing a painting that moves.

anthony powell

biography

Sandy Powell grew up in London, fascinated by fashion and the influential British couture designers working in the 1960s and '70s. She spent an important part of her teenage years going to the theater in London, including fringe productions well out of the theatrical mainstream. This early exposure motivated a start in designing costumes for innovative theater. Her work attracted the attention of director Derek Jarman,

sandy powell

known for edgy, nonconformist films. Among the films she made with Jarman are **Caravaggio** (1986), **The Last of England** (1987), and **Edward II** (1991). **Orlando** (Sally Potter, 1992,) was designed with a theatrical bravado rarely captured on film. A long collaboration with director Neil Jordan produced **The Crying Game** (1992), the seductive **Interview with a Vampire** (1994), Irish Republican history **Michael Collins** (1996), **The Butcher Boy** (1997), and resulted in a BAFTA nomination for **The End of the Affair** (1999). **The Wings of the Dove** (Iain Softley, 1997), received both Oscar and BAFTA nominations for Best Costume Design. In 1998, Powell achieved the remarkable accomplishment of receiving two Academy Award nominations in the same year, for **Velvet Goldmine** (Todd Haynes) and **Shakespeare in Love** (John Madden), for which she won the Oscar. She designed for Anand Tucker's **Hilary and Jackie** in the same year. 2002 was another prolific year, with **Far from Heaven**, a 1950s drama from Todd Haynes, and Martin Scorsese's mid-19th-century epic, **Gangs of New York**.

interview

For as far back as I can remember, I've loved clothes—from making dolls' outfits to dressing up. My mother used to make all my clothes, and taught me how to sew and use a machine while I was pretty young, so I started from there. I don't think I knew there was such a job as costume designer, although I was aware of fashion.

But by the time I was 13 or 14, I was going to the theater a lot. The person who influenced my decision to study costume and theater design was Lindsay Kemp, a dancer/choreographer who came to the fore in the 1960s and '70s and gave me my first job in the theater. He had an international company which performed highly avant-garde dance, and was also responsible for the whole Ziggy Stardust look for David Bowie. I saw one of his pieces, *Flowers*, in 1974 at the Roundhouse in London, and it changed my life. It was incredibly visual, with amazing sets, costumes, lighting, and makeup—and exactly the sort of work I wanted to be involved in. At the time he lived in Barcelona, as the theatrical style he created appealed much more to Europeans than to the stuffy old English audiences. When I told the teachers at art school that I wanted to work with him, they pooh-poohed me. "Oh, he's not interested in girls, and he lives abroad," they said. But I hunted him down. When I discovered he was teaching dance in London, I thought fine, I'll do a dance class with him. I did the class, and said to him at the end, "I absolutely love your work and would love to work with you one day." He just said, "Well, come to tea."

I showed him what little work I'd done in art school. I floundered around, wondering what I was doing there, and why I was doing it. But he said, "Oh, come and work with me one day." I took him at his word, and didn't go back to college that year. I thought, "Well, what's the worst that could happen?" They could turn around and say, "Go away," "No," or, "I'm not interested." But by Christmas he had called me, so I went off to Italy and designed two ballets, *Nijinsky* and *Façade*, with him. He said he gave me the job because he liked my purple hair. It's luck, but it's also about taking risks and chances.

The same thing happened with Derek Jarman. I had designed theater in London in the early 1980s—fringe theater, visual theater. It wasn't mainstream, and I used to make all the costumes myself. But I knew I wanted to do more, maybe in film. I'd always liked the look of certain directors' films—those of Visconti, Fellini, Ken Russell, and Derek Jarman—without knowing who the designers were. I managed to meet Derek, and invited him to come to the theater to see a show I'd designed, which he did. Again, he invited me back to tea. He took me under his wing and said, "OK, if you want to design films, maybe a good place to start would be in pop videos." So he introduced me to Tim Bevin and Sarah

Radcliffe who had a company called Aldabra that produced them at the time.

On the first pop video I designed with Derek, he showed me around the set, explaining everybody's role, and made me look through the camera. He was incredibly generous; he gave a lot of people their first break because he liked young people. With **Caravaggio**, the first film I worked on with him, I think the average age of the crew was 25. Honestly, my first film experience was hysterical, looking back on it. We didn't know what we were doing. We had no idea. We were making the costumes at the same time as shooting. Three weeks into the shoot, the continuity lady said, "Look, would it be at all possible to have somebody on the set?" I didn't even know you had to have somebody from the wardrobe department on the set. But I've learned from my mistakes. Derek was very similar to Lindsay in that he was also an artist and painter, he was good at imparting information, and he was generous with his knowledge. His starting points were always visual, and he'd show you paintings and books.

Neil Jordan is also visual, but in a totally different way. I can't explain how that collaboration works, it just does. He'll talk in his own way, and somehow I'll grasp it, and then present him with images. He then knows immediately, and he either responds or he doesn't. I'm lucky enough to have worked with directors who have given me freedom, and trusted me to do my own thing. But they also have to know what they want. Neil gives you direction—he'll want something specific that he's seen somewhere. He'll say things like, "That bit in that film, where she wears the hat for that scene, I'd like one of those." It's difficult to articulate the process we go through; it's more instinctive.

(1–4) **Shakespeare in Love**: Gwyneth Paltrow and Joseph Fiennes (1) are star-crossed lovers Viola de Lesseps and young William Shakespeare. The costumes for this 1998 multi-award-winning film are stunning Elizabethan period pieces, designed with a contemporary twist. The garments Powell designed, particularly for Dame Judi Dench's Queen Elizabeth (2) and Paltrow, are often elaborate—trimmed with beads, embroidery, or metal filigree—yet they are not stiff or unwielding. They look like clothes for real people, but are still imbued with a true sense of history, and make the character.

On **Felicia's Journey** (filmed in Cork, Ireland), an actor had to wear a flat cap, and I gave him one he didn't like. We were in the car driving home, and we passed a really ancient man of about 103 on a bicycle wearing the most horrible, dirty old cap. I stopped the car, ran out, and said, "Can I have your hat? Do you want this nice new one instead?" And I swapped hats. He couldn't believe it, he thought I was mad. He just stood there in astonishment in his brand new cap, then put it on and cycled off.

One of the first things I do after reading a script is to compile visual reference material that might be directly associated with a period. I start looking at the period specifically, whether it's paintings or photographs. Then I spread the net wider, and look at all sorts of images. I might look at fashion images, or photographs of people—a picture from a totally different period that has something about the face, or the way they're wearing their clothes, that interests me. I'll use that as a reference point for the character. I usually compile a huge book of images, and show that to a director. It's interesting to notice the bits they linger over, or the bits they respond to, as then you get a clue as to which direction to go in.

I think I differ from other designers in that I don't do drawings initially. I draw diagrams and sketches for my cutter. If I'm working with a cutter I haven't worked with before, or if I'm working things out for myself, I'll do scribbles, but nothing that I'd want to show anybody else. They're not beautiful costume drawings in color. If I do produce those, they're usually at the end of production, when they're needed for publicity. Hopefully, I'll meet the actors before I have to design anything, because I find it impossible to design a costume without knowing their shape or coloring. I'll start designing by looking for fabrics—I respond to the fabric—and work from there outward. After the designing, I'll start on the building/making process, and then the costume fittings. The most vital parts of the whole design process are the first and second fittings with the actor.

Martin Scorsese is incredibly visual and knowledgeable. Apart from waiting a year for **Gangs of New York** to start, I had about five months before we started shooting to prepare. Then we shot in sequence, which was great. That meant I was designing throughout the six-month shooting schedule, and we were still making costumes until about three weeks before the end of the shoot. It was a continual process. There were so many different sections and looks—it was relentless. I had to constantly keep inventing new costumes. We made the whole thing at the Cinecittà studios in Rome. We had two wardrobe workrooms and, of course, used independent tailors as well. It was epic, exciting, and a thrill to work with Scorsese. He was specific about costumes. Even though it's set around 1860, the world we were creating was not one that had been seen before. Scorsese wanted to create a world that was both believable and historically accurate, although a lot of it was also invented.

The scene which opens the movie is a battle scene, and the idea behind it was that it couldn't be placed in terms of period or location. The audience would think, "Where the hell is this? Where are we? What period is it?" It could have been post-apocalyptic, medieval, or futuristic—it was quite tricky. I looked a lot at various modern gangs, as well as those through the ages, noticing that most of them are distinguished by the colors they wear. We had to think of different identifying features for the gangs in the film, and chose things

(1–4) **Velvet Goldmine**: Powell's passion for costume design began in her teens, ignited by one performance at London's Roundhouse Theatre. She says a piece called *Flowers*, choreographed by Lindsay Kemp, who pioneered David Bowie's Ziggy Stardust look, "changed my life." Todd Haynes' 1998 pean to the Glam Rock era is heavily infused with Bowie-esque visual references. Ewan McGregor plays fictional American rock singer Curt Wild (3 right), whose show was very influential for his time, and Jonathan Rhys-Meyers plays a devotee of his style, Brian Slade (1–3 left). Sketch of Curt Wild costume (4).

2

3

4

(1–3) **Velvet Goldmine**: The film won a BAFTA and was nominated for an Oscar. It is one of Powell's favorite films she has worked on, and one of the most fun. Many of the costumes in the film have a homemade, customized look to them—Powell drew from her own experiences in 1974 in London (the year the film is set in), when she often made her own clothes.

like kerchiefs, hats, and hairstyles. There were certain specific gangs that had been written about and documented, but others' looks I had to invent.

Sally Potter was quite specific about how she wanted **Orlando** to look, and her references. It was always going to be a highly stylized piece, and it was early in my film career. I think the look was influenced by Derek Jarman and Peter Greenaway films, combined with Sally's, and my ideas. A lot of the costumes were made by people who worked in the theater, and that theatricality comes across. Just before designing **Orlando**, I had designed the play *Edward II* at Stratford-upon-Avon, and the costume maker from that production made the costumes for Tilda Swinton in **Orlando**. After that, I designed **The Crying Game**. The main task on that film was to try and make the audience believe the actor Jaye Davidson was a woman, which was pretty interesting. But when I first met him, he did look extraordinary, and all I had to do was provide some extra padding. It was one of those low budget, scramble around, do-what-you-can-on-the-money movies. Gender things, I keep doing those, don't I? Boys disguised as girls, and girls disguised as boys, and the same with **Shakespeare in Love**, and **Velvet Goldmine**.

I have favorites for different reasons. **Velvet Goldmine** was one of the most fun experiences, and the most personal, really. It was set in 1974 in London—my time of awakening. I was really into the clothes and the lifestyle. I made my own clothes, tried to be cool, and wished I had more money so that I could buy fabulous ones. It was a homemade look, which I incorporated into my design for the film. A lot of people criticized it, saying it wasn't really like that. But I thought, "Well, maybe it wasn't in your world, but it was in mine."

The difficulty with contemporary films is that everybody has an opinion about the clothes. If you're designing a period film, the actor and the director don't necessarily know about the details. You're the person who knows more than anybody else. I think actors find it harder to think of themselves in character in contemporary films. I'd find it difficult. I'd think, "Oh, I never wear that color," or, "I don't wear that shape, ever." You want to remind actors, "Well, you don't, but maybe your character does." It's the same when you're trying to create somebody who has bad taste—that's always quite difficult. Nobody wants to look horrible. In real life, people can put clothes together that look really ugly. That's quite difficult to recreate in a contemporary film. Your natural instinct is to make people look fabulous, but most people don't. In some ways, 20th-century costumes are easier.

With a period, such as Elizabethan, people in production are often worried that the modern audience isn't going to relate to the characters. You have to do something to give it contemporary appeal. For example, I designed the jacket for Joseph Fiennes, who played Shakespeare, as if it was a modern-day leather jacket, rather than a doublet.

It was important, when designing real events, such as the Irish Civil War in **Michael Collins**, to connect to the characters. You want the audience to be involved with the protagonist and his cause—you don't want to be distracted by his clothes. If a contemporary film is designed well, you don't notice what they're wearing. That's the unfortunate thing about costume designers who are good at designing contemporary films. They don't get recognized, no-one says, "Wow, the costumes were great." You don't notice them, which means they were great.

1

2

3

(1–7) **Orlando**: Based on English novelist Virginia Woolf's modernist novel written in 1928, this film is a commentary on representations of history and sexuality throughout 400 years of English history. It stars Tilda Swinton as Orlando; she goes from being a male courtier who curries favor in the 16th-century court of the Virgin Queen Elizabeth, played by Terence Stamp (4, 7 over page), to the polite literary salons of the 1750s, where she has evolved into a woman, to the Victorian era, and finally to the 20th century. In Powell's words, this was a "highly stylized" film; many of the costumes were made by people in the theater, but this clearly translated into an appeal for cinema audiences. The film received a nomination for both an Oscar and a BAFTA award.

6

4

5

Hilary and Jackie is about the cellist Jacqueline du Pré, who died in the 1980s of multiple sclerosis. The script was based on a book written by her sister, who dared to suggest that Jacqueline wasn't a particularly pleasant person. The director, Anand Tucker, wanted it to be more than a straightforward biography, with a look and a feel. It's almost a contemporary film, but not quite. It runs from du Pré's childhood in the 1950s through to when she dies. There's an extraordinary quality to it, as the style isn't totally naturalistic. I looked at lots of photographic references, but didn't duplicate what people were wearing. I don't think, "Right, let's recreate that outfit," because it doesn't really matter—it's getting the feeling right that counts.

The most important thing about costume design, and the most exciting part, is helping to create a character and contribute to a story. It shouldn't be about making somebody look fabulous. It's not about how glamorous, or sexy, or wonderful somebody looks. I actually like working on low-budget pictures, because I find them challenging. It means you can't rely on getting a costume made overnight, or that you can pay someone double to get it done quickly. You have to think about how you can get the most out of what you are working with, and it forces you to be creative. I chose to do the films that are interesting, but the interesting films don't get the big budgets, or make the big money.

Gowns by Adrian in **Marie Antoinette** (W. S. Van Dyke, 1938) inspired Bob Ringwood to become a costume designer. But rather than designing conventionally elegant dresses from centuries past, Ringwood's innovative work has challenged him to invent a new look for fantasy and futuristic films. In many ways Ringwood has changed the way futuristic films are costumed, not only by his groundbreaking work in the field of

bob ringwood

specialized costume construction, but by the texture and silhouette of his own designs, he has redefined the way the future looks. He won the Academy of Science Fiction, Fantasy, and Horror Films' Saturn Award for Best Costume for his first two feature films, John Boorman's **Excalibur** (1981) and David Lynch's **Dune** (1984). Because of the very unusual, often unrealistic settings of Ringwood's movies, he has had the opportunity to create new worlds, such as those in **Dune** and the **Batman** series (Tim Burton, 1989, 1992 and Joel Schumacher, 1995). For Steven Spielberg's **Artificial Intelligence: AI** (2001) Ringwood succeeded in designing highly plausible costumes for "sometime in the near future." Although the use of plastic and other "alternative fabrics" may be a calling card for the designer, he has also excelled in every other costume genre. His meticulous period costumes for Steven Spielberg's **Empire of the Sun** (1987) earned him an Academy Award nomination, and his understanding of costume's narrative force is seen in his more recent **The Time Machine** (Simon Wells, 2002).

interview

I wanted to be a painter. I was at art school in London in the 1960s when the focus was on abstract art. This didn't suit me, as I wanted to paint figurative pictures, and I got very disillusioned. A friend told me that my paintings were theatrical and that I should go to the Central Saint Martin's School of Art and Design in London to study theater design. I took her advice. Part of the coursework involved studying costumes with Margaret Woodward, who wrote the two definitive books on costume construction and cutting with Norah Waugh. In the morning she would show us a painting, give us a lot of muslin, a stand, some pins, and some scissors, and say, "Cut that costume on the stand." Later she would come back, cut it properly, and show us what was wrong with what we'd done. It was a brilliant way of learning, because you made your own mistakes. We would lay our patterns next to those that she had cut to see the difference. I learned a huge amount about costume cutting and how to make clothes from her.

After I left I designed in the theater for about ten years. I was a set designer, and had to design the costumes for the productions as well. I worked on about 150 productions and, out of necessity, I got very good at designing clothes. My friend David Walker was designing the costumes for George Cukor's **The Corn is Green**, starring Katharine Hepburn. He asked me to assist him because he liked the way I worked. I was excited to be working on a film for the first time, and being over-efficient, I called Katharine Hepburn at 9:30pm the night that she arrived from New York, to let her know that she had a fitting at 8:30 in the morning. She had just flown in, so she was very tired. There was an endless 30-second pause, and she said, "Bob Ringwood, I am not an imbecile," and hung up. However, she was charming the next morning and said, "Take no notice. Let's go and have coffee." David became ill, and Ms Hepburn insisted that the studio hire me to replace him, although I'd never been near a film set. She kindly wrote a letter to the union and I finished the movie. I would have dinner with George Cukor three nights a week and dinner with Katharine Hepburn the other two. I learned so much about the film industry from them and how to not be trampled on. At the end of each evening, Katharine would open an old diary and she'd say, "Let's read 1933 today. Summer 1933." And she'd sit and read about the death of Spencer Tracy or the making of **The Philadelphia Story.** The whole experience was like a dream—a crash course in moviemaking from two of its greatest experts. David recommended me to director John Boorman to design for **Excalibur**. From there, my theater career stopped abruptly and almost overnight, I became a costume designer in movies.

I loved working on **Dune**. The director, David Lynch, gave me a great deal of freedom. I pulled out all the stops and used all sorts of obscure references. I had read the *Dune* series, by Frank Herbert, and I saw in it many elaborate, fantasy images. Anthony Masters, the production designer of Kubrick's **2001: A Space Odyssey**, was also the production designer on **Dune**. He had been working on the film for about nine months when I arrived. The sets were full of silver tubes and high-tech rubber tubing—very sci-fi. I didn't think that look was right for the project. I had envisioned a film with a much richer, more sophisticated texture. The producers, David and Raffaella De Laurentiis, agreed, and the set designs were scrapped. Tony Masters behaved like an angel. He said, "Well, let's see what visual styles we can come up with." I suggested that we model the film on the work of Otto Wagner, the Viennese architect. Wagner's book became a bible for the visual style of the film; very elaborate, but restrained at the same time. When designing a film, I often use what seem to be irrelevant references. For example, if I'm designing a chiffon dress for Michelle Pfeiffer, I might bring in pictures of the inside of a refrigerator door, because there's something about the quality of that photograph that I feel the dress must capture. It's an odd way of working. Sometimes directors just don't understand it, but it's great when they do.

Usually my relationships with directors are good, but sometimes you do have to guide the director. When I designed the costumes for **Demolition Man**, the director, Marco Brambilla, seemed to be somewhat confused about the look of the film. After it was made, he said, "At the time I had no idea what you were talking about. Now I see it finished, and if you hadn't pushed it through we would have had nothing."

When I design costumes I like them to be as real as they can be, to create their own reality, even if they're fantasy

(1–3) **Dune**: Raffaella De Laurentiis commissioned Ringwood to design for **Dune** after seeing one of the dresses he had designed for **Excalibur**, only his second film. Before going to work in Hollywood for the first time, Ringwood read the *Dune* books (written by Frank Herbert) and came up with a singular vision for the style of the film that totally went against the ideas of production designer Anthony Masters. Still (2) and prototype (3) of one of the unmistakable **Dune** costumes.

1

The homemade look of Catwoman's costume inspired an extra scene in **Batman Returns** in which Catwoman (Michelle Pfeiffer) makes her outfit (1). Ringwood's costume for the eponymous hero of **Batman** was a sleek, tight-fitting black rubber bodysuit (2), but he applied what he calls a "film noir Rockwell" feel to the other characters in the film, using hats and clothing reminiscent of 1930s and '40s films. Sketch of The Joker's colorful costume from **Batman** (3), and sketch of The Penguin, the reincarnation of The Joker, in **Batman Returns** (4). Ringwood designed costumes for the first three of the Batman saga, bowing out on the fourth, Joel Schumacher's **Batman and Robin**.

2

3

4

spacesuits. For Spielberg's **Empire of the Sun**, I tried to make the costumes look real but add a slightly heightened edge; a Hollywood edge. I was in awe of Steven Spielberg, but I enjoyed working with him on the movie. I did an enormous amount of research for that film, and I tried to recreate the images of the photographs I saw. I didn't realize that Spielberg was quite as romantic and sentimental as he is, and I think that my designs may have been a bit raw for the visual style of the film, but they did help give it bite. We only had six and a half weeks to get ready for **Empire of the Sun**, and we had about 10,000 costumes to make. 6,000 of them were made in China, which was the most economical place to do them. We paid nine dollars for full outfits—fully quilted, hand-dyed trousers, jackets, vests, hats, and shoes. They were beautiful. They were made better than you could have made them anywhere else because of the amazing Chinese work ethic. They made thousands of dragon-knot buttons by hand, and every one was perfect.

The **Batman** films were an extraordinary experience. (Unfortunately, they earned me the reputation as the designer of rubberwear.) I had spent about ten minutes meeting with Tim Burton when he hired me to design the costumes for **Batman**. I think he'd never met anyone who talks as much as I do. I have to confess I'd never seen a *Batman* comic. I read about six of the early comics before I designed the movie. For me, the goal was to get the essence in my head and go for it. I expected a big, superhero type of actor to play Batman, but they cast Michael Keaton. Michael is not overly muscular, so we had to turn him into a superhero, hence the rubber muscle suit. I was aiming to make his costume a "super-costume" so that it was almost unbelievable. I wanted Batman to be bigger than reality on the screen, and by the time we filmed the third

one with Val Kilmer, we had perfected the batsuit. It couldn't get sleeker. Think about it, you're creating a man who puts on a black suit, with long, pointed ears, and goes out in the night. It's a bizarre thing to make into a reality, but I think we pulled it off very well. The costume budget on the first **Batman** was $100,000—nothing for a movie of this type. I called every favor in, and even with very little money, we managed to make the film look pretty good. We spent a lot more on the sequels.

On **Batman Returns**, Tim Burton had a great deal of influence on the way that Catwoman's costume looked. He wanted it to be unexpected, strange, and offbeat. I showed Tim a photograph of a sculpture of a woman's head that looked like she'd had her skin sewn on. It was a strong image, and Tim, costume designer Mary Vogt and I decided her costume would be black with white stitching, and it would look homemade. Tim invented a scene where Catwoman makes her costume from her patent leather slicker-style raincoat to justify the costume.

If you're a good costume designer, you're helping to invent the character for the actor. The actor invents the personality, but you invent the character together. The costume designer and the set designer conceive the world in which the characters live. If you're clever, you give the actor a character to put on. You can be very helpful, or very destructive to the actor, depending on how well you do that job. For example, when Piero Tosi (probably the best film costume designer there is) designs period films, he doesn't just design a lot of nondescript historical clothes. He designs an absolutely specific character for each person, and he carries it out to perfection. His clothes for **The Leopard** are as near to heaven as you can get.

bob ringwood

costume design

1

2

3

4

(1–9) **Artificial Intelligence: AI**: Sketch and costume for David, the robotic boy played by Haley Joel Osment (1, 4). Jude Law (Gigolo Joe) in costume in wardrobe, and production still featuring the same suit (2–3). Detailed sketch (5) and close-up of a biker hound's costume (6). Test shots of stunning female costumes (7–9). Although much of the film was based in modern interiors and featured modern costumes, Ringwood designed and made almost everything. "I think, unless you make the clothes, you're not really designing it—you're actually styling it."

5

BIKER HOUND

6

7

8

9

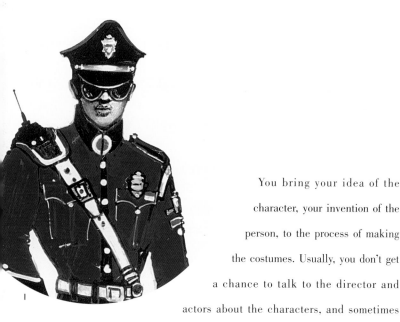

You bring your idea of the character, your invention of the person, to the process of making the costumes. Usually, you don't get a chance to talk to the director and actors about the characters, and sometimes you're designing costumes before the movie has been cast. I try to manufacture almost everything for the films I design. For instance, for **Artificial Intelligence: AI**, although the costumes were modern clothes worn by an ordinary, middle-class family, sometime maybe 30 years in the future, and we could have bought the clothes, I chose to make them, because you get exactly the look you want.

Each costume has to be made by the person who is most appropriate to make it. To me it's like casting a film. I find references and photographs and then the sketch artist, and I start drawing. I look for the fabrics and then decide who should make the costume. When a costume goes wrong, it is generally because you've chosen the wrong person to make it.

The designer really has to know what he's doing, but he also needs to know who to hire. Although I use the same makers that a lot of other people in Hollywood use, I try to understand their strengths and weaknesses, that way you get the best work out of them. There's no point in having a brilliant design for a costume if you can't get it made. I'm always saying to producers, directors, and actors: "Remember, you cannot wear a drawing." A drawing is just the beginning. It's the creation of it, the fabrics, and the cut, that count. The cut is everything. For example, Anthony Powell's clothes for **Tess** were brilliant, because of their cut. A good cutter is the lifeblood of a costume designer.

All this effort is to create a character for the screen. If a costume designer gets too involved with fashion, he's not doing the job properly. The designer should focus on each particular character. Actors have told me that when they put on the costumes for the first time in fittings, they start to see the person they're creating in their head. Sometimes an actor's understanding of a character will be slightly different from the designer's, and you must design the costume to reflect and support the performer's conception.

When I design a film, I like to set up a factory and have all the cutters and seamstresses in my own workshop. I like to have complete and total control over everything that's made, every button and boot. A lot of producers won't hire me because they think that's an expensive method, but it's no more expensive than manufacturing all over the city and paying the overhead and profits. When everyone is working with me on the premises, they get immediate answers to their questions. I try to work like designers did in the old Hollywood studios. I make sure the actors come to the wardrobe department. I won't fit the costumes in trailers, with bad light and no space.

When I design costumes I think about how fabric reflects light, and how the camera sees them. The cinema interests me because the camera is a recorder of light. Unfortunately, modern technology with its digital cameras is not a great friend of light. To me, digital cameras make contemporary films look flat and dead; too colorful. The beauty of our work is the interaction of the mind and hands that create it. I love the coarseness and texture of handwork, and to feel the way a thing is made. David Lynch said to me on **Dune**, "I don't want any of these dresses to have any wrinkles." I said, "The

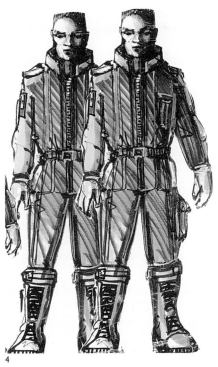

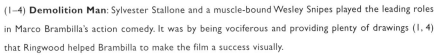
(1–4) **Demolition Man**: Sylvester Stallone and a muscle-bound Wesley Snipes played the leading roles in Marco Brambilla's action comedy. It was by being vociferous and providing plenty of drawings (1, 4) that Ringwood helped Brambilla to make the film a success visually.

(1–2) **Excalibur**: Bob Ringwood was brought on to the film six weeks into preparation, following a telephone call to his mentor David Walker from director John Boorman.

2

beauty of the way silk wrinkles, is what silk is all about." The American dream is to have a perfect wrinkle-free costume. After all, they invented the stay-pressed trouser—horrible!

Curiosity is the most important quality in a costume designer: curiosity about why the buttons are placed the way they are, why the seams are sewn like that, why the fabric is so rough on one side, and why people powdered their faces. I asked a 90-year-old friend why he is so youthful and successful, and he said, "Because I'm curious." Curiosity drives you to stay that extra hour and add that last little change to the costume, but the real secret is to know just when to stop work on the costume so it stays fresh.

The way costumes are designed has changed. When I design costumes for a period picture, I read novels of the period. The stories of the time make the period start to come alive for me. Young designers are not as interested in recreating historic periods as we older costume designers are. They want to invent more, and why not? But I do see the quality of craftsmanship disappearing. Hats, for instance, are not as well-made as they once were. A bad hat sits on the head. A good hat becomes part of the head; then the viewer isn't aware of the hat and only perceives the character.

If you want to be a costume designer, I suggest studying historical clothing and learning how clothes are cut and sewn. Spend years of your life learning how things were put together in the past, and understand why they were put together that way. It is this knowledge that makes Tosi such a genius. He knows exactly when a piece of braid should be cobbled on or left loose and hanging, or when a piece of braid should be tied down and tightly sewn, how a costume should have air and lightness. If you're designing contemporary costumes, really look at modern clothes—not in a magazine, as a magazine photograph is computerized and has nothing to do with real clothes. Really look at those clothes and analyze them. Above all, be curious.

biography

With an amazing career spanning 40 years, the indefatigable Ann Roth has designed the costumes for an astounding 90 films, and the quality of her work has never varied. Her approach to her work is always thorough and truthful, creating a devoted following of actors and directors. Roth was raised in Pennsylvania Dutch country, and she studied set and costume design in Pittsburgh before working in the theater in New

ann roth

York. Her thoughtful, deliberate approach to creating each character has been demonstrated time and again. She is a master storyteller, and the consummate film artist and collaborator. She has worked with many distinguished directors, including John Schlesinger: **Midnight Cowboy** (1969); **The Day of the Locust** (1975); and **Pacific Heights** (1990), and Alan J. Pakula: **Klute** (1971); **Rollover** (1981); and **Consenting Adults** (1992). Roth has also collaborated with Hal Ashby: **Coming Home** (1978); **Second-Hand Hearts** (1981), Brian De Palma: **Dressed to Kill** (1980); **Blow Out** (1981); and **The Bonfire of the Vanities** (1990), and Sidney Lumet: **The Morning After** (1986) and **Q & A** (1990). Roth's costumes won an Academy Award for Anthony Minghella's **The English Patient** (1996) and a nomination for his **The Talented Mr. Ripley** (1999). Her best-known professional relationship, however, is with Mike Nichols, with whom she worked on ten films, including **Silkwood** (1983), **Postcards from the Edge** (1990), and **The Birdcage** (1996). Roth also designed the costumes for Stephen Daldry's provocative **The Hours** (2002).

interview

I've been designing costumes for movies for nearly 40 years, and I'm still excited by every new project. A recent film I have worked on is **Cold Mountain**, directed by Anthony Minghella, shot in Romania. This is my third collaboration with Anthony. The first film that I designed for him was **The English Patient**, and we worked together again on **The Talented Mr. Ripley**. I appreciate the comfort that comes from working with the same director on several projects. By the third experience, the language between the designer and the director has become simplified. We understand each other's process and trust each other's work. At that point, we can just enjoy the gratification of the artistic journey. I have enjoyed working on **Cold Mountain**, in part because of my relationship with Anthony, and also because it enabled me to spend time in Rome working with the golden hands of the Tirelli Costume Company. Many of the films that I've been asked to design costumes for lately are financed multinationally. Actors live all over the world, and in this instance, it was important to find a center strategic to Romania. The film is set during the civil war in 1860, and the artists and craftsmen at Tirelli have been very enthusiastic in exploring a new aspect of the mid-19th century.

I grew up in the farmland of Pennsylvania Dutch country, and I wanted to be a painter. I studied design in Pittsburgh and was accepted into the art program at Carnegie Mellon. I found the art world very exciting. Andy Warhol was at Carnegie. It was a very small school, and I had a fabulous experience. I got involved in painting scenery when I was a teenager, and that is how I was introduced to life backstage. I had painted sets for the Pittsburgh Opera House and had moved on to the Buck's County Playhouse when I met Irene Sharaff. She told me that I should come see her when I graduated. I took her at her word and went to see her in California.

Brigadoon was the first film I worked on as Irene's assistant. I worked for her on five movies and five Broadway musicals. Irene Sharaff was a wonderful mentor—she was absolutely driven. Her research for a project was tireless, and she would seek out the most obscure sources. She made her drawings on the finest watercolor paper, and her paints were the most difficult to obtain. Irene had work habits and discipline that I had never imagined. She was tall and beautiful and frightening, and all of the wardrobe staff—every milliner, tailor, glovemaker, feather-dyer, shoemaker, jewelry-maker, lacemaker, woolweaver, and leatherworker—wanted to work for her and amaze her. Because when he impressed Irene Sharaff, that craftsman was living and operating at his peak. He was reminded of the reason he was doing what he did. From her, I learnt the decorum and procedure necessary in the fitting room, but perhaps more importantly, I learnt how to generate excitement—like that of the Tirelli staff.

I was an inventive shopper and I worked for her on five film musicals and five Broadway musicals. She and I used to discuss who made the better assistant: the person who wants to be a costume designer, or the person who wants to be an assistant? It seems to me now that the person who wants to be an assistant is the best person for the job. The good assistant needs to be able to fill in the crevices of the designer's work. Like many costume designers, I am probably hard to assist, because I have a hard time delegating. I have to go to every fitting—I have to fit the secretary, and the housekeeper, and the policeman.

My other mentor was Miles White. I worked for Mr White on **Around the World in Eighty Days**, the *Ringling Brothers Circus*, and quite a few Broadway musicals and plays. He was completely different from Ms Sharaff. His sense of humor and his eagle eye in the fitting room were awesome. He was a born designer. He could take a bedspread and turn it into something sensational. I knew him and Ms Sharaff well, and in truth I would say that costumes were their lives and obsession. I don't remember either of them engaging in any activity other than creating costumes.

These days designers will often set up their own workrooms, and if you have the money it's a wonderful way to create costumes. When I started in Hollywood, the studios still had one or two designers under contract and active wardrobe workrooms. Helen Rose and Walter Plunkett were at MGM. There was a reverence for the craft of costume making then. Today you don't see people developing expertise in aging costumes, hand-stitching, crocheting, or embroidering. We no longer care about whether someone can make a bonnet. Costume designing has changed quite a bit in that way.

(1–3) **Places in the Heart**: Roth was nominated for an Academy Award for her designs for this emotional story of a sheriff's widow (Sally Field) fending for herself in a small Texas town in 1935. Sketch of costumes for the cotton pickers featured in the film (2).

cotton pickers.
Places in the Heart.

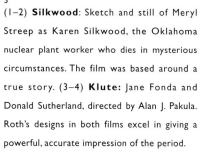

(1–2) **Silkwood**: Sketch and still of Meryl Streep as Karen Silkwood, the Oklahoma nuclear plant worker who dies in mysterious circumstances. The film was based around a true story. (3–4) **Klute:** Jane Fonda and Donald Sutherland, directed by Alan J. Pakula. Roth's designs in both films excel in giving a powerful, accurate impression of the period.

The amount of research I do for a film is extensive. When I was in school in Pittsburgh many years ago, I was determined to learn to design scenery and costumes. One of my first jobs was **Ring Around the Moon**, and after I had designed it, drawn it, completed the working drawings and paint elevations, the director asked me about my research. He was an older, very experienced, and highly cultured man. My research had been pedestrian, unimaginative. I simply wasn't curious enough. He knew it, and I learnt the most significant lesson of designing essentials. Research, and more research. Now I probably spend way too much time gathering source material, but it's a part of my process. For **Cold Mountain**, I've collected countless books on the civil war in Romania.

My research process on a contemporary film is different. When I was preparing to design costumes for **Working Girl**, I went to the World Trade Center and watched people getting on and off the Staten Island Ferry. Staten Island, where the heroine was raised, actually had its own style and flavor. It was the late 1980s, and the hairstyles were very exciting to me; all the women had big hair. They carried their office shoes in their bags, and they were sexy. It was the best way I could think of to get a genuine sense of how to dress the principal actress. Being observant is an important quality in a costume designer. The designer should look and search out and be in awe. He should notice the hair, the girdle, the no girdle, the carriage of the body, the stains, the starch, the balance. He should want to share his way of looking—his interpretation. A costume designer should draw as much as possible, and should enjoy making quick sketches for himself.

Being a costume designer has shaped me into a tirelessly curious person. Art exhibits, museums, primitive cultures, jewelry, pottery, sculpture, rugs, north-east Russia, south-east China, outside art, the 18th century: all of them are fascinating and have something to teach. This curiosity is a blessing. Whether it's a modern political piece (yet another Burberry raincoat) or a domestic children's piece, a **Working Girl** or a **Mambo Kings**, it allows me to approach a new script with excitement.

Often I'll find a production designer or a director who is attracted to the same visuals as I am, and that probably means that we have similar sensibilities. It's true that certain senses of humor gravitate towards one another. Though I admire many directors and production designers enormously, we often do not have a similar sense of humor and it may be difficult for us to work well together.

I had a wonderful relationship with John Schlesinger. Of the films we did together, **Midnight Cowboy** and **The Day of the Locust** were my favorite experiences. John and I tend to dissect and analyze the characters endlessly, and our mutual interest in the story behind the characters makes us a good team. I like to really imagine a character's life: his income, where he sleeps, who does his laundry, where do his clothes land when he takes them off at night—on the floor? Is he a neat person? What is his visual, sensual fantasy of himself? What does he read? Does he read? And so on. This is my search when creating a character, and it was great to work with a director who is so supportive of the process. Anthony Minghella also likes to discuss characters with me.

I only designed one movie for Gus Van Sant, which was **Finding Forrester**, but I'd love to work with him again. We became quite close, and I felt that he had faith in me. I told

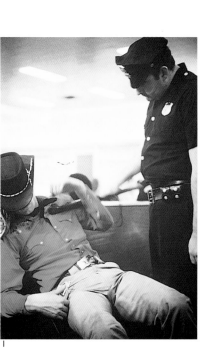

1

2

3

4

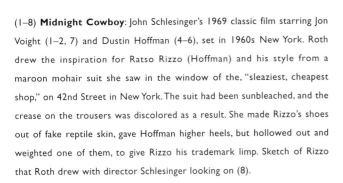

(1–8) **Midnight Cowboy**: John Schlesinger's 1969 classic film starring Jon Voight (1–2, 7) and Dustin Hoffman (4–6), set in 1960s New York. Roth drew the inspiration for Ratso Rizzo (Hoffman) and his style from a maroon mohair suit she saw in the window of the, "sleaziest, cheapest shop," on 42nd Street in New York. The suit had been sunbleached, and the crease on the trousers was discolored as a result. She made Rizzo's shoes out of fake reptile skin, gave Hoffman higher heels, but hollowed out and weighted one of them, to give Rizzo his trademark limp. Sketch of Rizzo that Roth drew with director Schlesinger looking on (8).

(1–3) **Working Girl**: Roth drew from time spent people-watching when preparing for this film. She watched women getting on and off the Staten Island Ferry—the boat that ferries commuters from Staten Island to Manhattan—and observed their hairstyles, shoes, etc. From this research, she assembled a picture of working girls in the 1980s. The costumes in the film are a triumph of contemporary costume design, and perfectly sum up the zeitgeist of the time.

him he could come to my first fitting with Sean Connery, but that he couldn't speak unless I gave him permission. My intent was to just play around at this fitting. He could actually participate in the second one.

My relationship with Mike Nichols, on the other hand, goes back 36 years. Mike is constantly reinventing as he works; he is always looking for something new. He is very demanding, very confident, and he never mentally tires. Mike is very patient with me. Because our association is so long, I find myself often overreaching. I don't want our way of working to get stale. We will usually envision the script and the characters in the same way, only he's smarter. In fact, he's the smartest. He would deny it, but we both have a love of improvisation. He likes a scene to take shape from an improvisational rehearsal, and I like accidental problem-solving. We were working together on **Heartburn**, and he pointed to a dress and said to me, "You told me this dress was going to be pink." I said, "I never, never said pink, but okay, it will be pink." I painted the dress with Pepto Bismol (a digestive remedy). It looked good.

Ratso Rizzo, Dustin Hoffman's character in **Midnight Cowboy**, inhabited the world of 42nd Street in New York. Marcello Mastroianni's name was prominent in movie marquees along the street at that time, and I imagined that Ratso adopted his visual identity from this glamorous Italian's. In fashioning himself to approximate Mastroianni, he found himself a white ensemble (trousers from a sale table in front of Port Authority bus terminal and a discarded white prom dinner jacket). Ratso bathes in public restrooms and desperately tries to maintain his vanity. When Dustin and I stood in the fitting room, we experienced the process where the reflected image slowly becomes another person; not the actor, but the faint beginnings of a foreign body. The actor begins to feel free, and the creative energy between the two of us starts to flow. Soon we see Ratso, his crippled leg, his dirty fingernails, his high school ring. This moment is the high point of costume design. If the actor and I have reached this point together, he knows that his journey is safe and will be aided by me. If he's happy, the director's happy.

The key to designing costumes successfully is to be brave in giving the actor the foreign element that frees him from himself. The minute you do something to transform somebody —when you cut their fingernails or give them dreadlocks— you're liberating them. Every actor I know is very insecure when he takes his clothes off. I want to dress him and have him open his eyes and sense the birth of a character. I don't know any costume designer who doesn't want to help an actor. Placing the actor in the literature, freeing the actor to take risks, is one of the most rewarding parts of the job. It's what I like to do.

biography

The designs of Piero Tosi are cited as an inspiration by virtually every costume designer of the modern era. Tosi's career bridges four decades of films featuring his exceptional sense of drama and realism. For Tosi, costume design is centered on creating authentic characters to serve the narrative—and he has been immensely successful. **Bellissima** (1951), Tosi's first feature film, was directed by Luchino

piero tosi

Visconti. Their collaboration continued for 25 years. Tosi was nominated for an Academy Award for three Visconti films— **The Leopard (Il Gattopardo,** 1963), **Death in Venice (Morte a Venezia,** 1971), which won a BAFTA award for Best Costumes, and **Ludwig** (1972). He worked with Mauro Bolognini on **You're On Your Own (Arrangiatevi!,** 1959), **The Lovemakers (La Viaccia,** 1962), **Arabella** (1967), and **The True Story of Camille (La Dame aux Camélias,** 1980), and with Vittorio De Sica on **Yesterday, Today and Tomorrow (Ieri, Oggi, Domani,** 1963), and **After the Fox (Caccia all Volpe,** 1966). For Federico Fellini, Tosi designed the costumes for the *Toby Dammit* segment of **Spirits of the Dead (Tre Passi Nel Delirio,** 1968). Tosi began working with Liliana Cavani in the mid-1970s when he designed costumes for **The Night Porter (Il Portiere di Notte,** 1974), and continued this relationship for the next decade. He was nominated for another Academy Award for **Birds of a Feather (La Cage aux Folles,** Edouard Molinaro, 1978) and Franco Zeffirelli's **La Traviata** (1982). In 2002 Tosi received the Costume Designers Guild inaugural President's Award.

interview

For me, designing costumes is a passion that arose from my childhood. I was enchanted by the photos that hung on the pegs in the newsagents' kiosks. These were the images of divas—a galaxy of visual images uncommon on the streets during the war, because the only thing you saw then was the misery and the squalor on people's faces. These new images full of light—the divine faces of Marlene Dietrich and Greta Garbo—made a great impression upon me.

In 1948, there was a huge festival in Florence, the Florentine May. All the great directors came from all over the world, such as Jean Renoir and Luchino Visconti. Franco Zeffirelli (my schoolmate), who had worked as Visconti's production designer on **Troilus and Cressida**, was also there. I went to see him and secretly showed him some of the work I had done at college. The old cinema traditions from the 1920s were still prevalent then, and I was rebelling against them. I thought that production design should be approached in a different way, with more research and more authenticity, closer to the actual script and the nuances of the script's ambience.

Thanks to the involvement of Zeffirelli, I was called to the set of **Bellissima**, a Suso Cecchi D'Amico and Visconti project with Anna Magnani. All the costumes had to be done in a very short space of time. I was very young and naïve, but exploding with ideas. I didn't then communicate directly with Visconti, but spoke instead with the writer Franco Rosi and with Zeffirelli. I was only 22, with a very young boy's face and didn't inspire confidence.

Visconti didn't want to have tailor-made costumes, but rather, ragged clothes taken from real people. It was the time of neo-realism, when everything was to be authentic using real clothes, so I went in search of this look. Luckily at the time, the public was fascinated with cinema, and Magnani was revered as a myth. As soon as I would mention to the people on the street the two words "Magnani" and "cinema," people were ready to take their clothes off and give them to me! Which made everything much easier for me. I know I was very cheeky, but it worked. I once followed a lady in the Nerioni market, and when she turned to face me I told her that we were making a Visconti film with Anna Magnani, and I would be very interested in her suit. She replied that that would be impossible because it was ugly and old, but I convinced her that it was perfect, so she gave it to me and she gave me everything else, even the shopping bag. This way of making a costume, to take real things and dress the actors in a way that would be suited to their personalities, made me understand how false a costume can be when it's made from scratch with all the right measurements. Such an outfit can give a very contrived image. Using real clothes, the actors became alive, and immediately I felt a warmth and a sense of authenticity. This was a great lesson, and helped me continue to create modern costumes that embodied a feeling of truth. I learned more on **Bellissima** than I had at any school.

In school they taught how to draw costume sketches, how to make them lovely and pleasing to the eye. But these drawings were only beautiful, descriptive images, and I found such lessons insignificant. It was presumptuous, but when I worked with Visconti, I began to think that maybe I was right. This type of decorative sketch doesn't mean much. Sometimes the producer and the director ask for traditional costume sketches, and you're obliged to do them. As far as I'm concerned, though, they can stick them on the wall, because I have never needed them. I always did all sorts of preparation for Visconti—I was known as the costume and production designer who paid attention to every detail—and then, on the eve of filming, everything would be changed. One morning Visconti came out from the bathroom with all his clothes on the floor and he said, "This is the way I would create a natural scene." I would go to work with his vision in my mind.

I try to choose a director who gives me a good script, one that I can feel passionate about. It's easy to work with a director of quality who you respect. What's interesting is that you have to be a little more careful with a director who doesn't ask for much. Visconti, for example, had very precise ideas. He would propose a scene, set it in a certain period and where he would like to shoot it, and then he liked me to do the research, and I would provide him with all the necessary information. When we did **The Leopard**, for example, all the decorations and the photo shoots of the family furniture and family clothes were prepared in Sicily.

2

3

4

Tosi had a long working relationship with director Luchino Visconti. **Ludwig** (1–4) and **The Leopard** (5) are two of his finest films. The designer says that working for Visconti felt more like "a moral duty" than working for other directors, and he always said yes to designing one of his films. "It was very hard work, but I persevered." **Ludwig** charts the history of Ludwig II, King of Bavaria between 1864 and 1886, and was made in 1972 starring Helmut Berger and Romy Schneider. **The Leopard** was an earlier film (1963), set and filmed in Sicily. Although Visconti wanted the look of a true Sicilian woman for the lead role of Angelica Sedara, he cast Claudia Cardinale, the Italian-born actress, and Tosi worked hard to make her look less groomed and less refined.

5

1

2

3

(1–4) **The Damned:** Another collaboration with Visconti, this 1969 film charts the history of the Von Essenbeck family, wealthy owners of an industrial empire during the rise of the Third Reich in Germany. Each character symbolizes an aspect of German history in this powerful insight into Nazism and its consequences. Tosi had done seven months of preparation with actress Vanessa Regole when the Swedish actress Ingrid Thulin took over. As Thulin's look was more typical of the 1950s, Tosi kept working on her hair, makeup and costumes to turn her into a 1930s woman.

4

When I went to work with Pasolini, he looked at things in a different way from Visconti—we fought, and he looked at me suspiciously and used to say to me, "You want to give me Visconti's world," so it became a terrible battle. It wasn't that way with Fellini. His was a different world, and I had to adapt to his world. Fellini was always uncertain, he would wait for divine intervention. If he wasn't ready to begin filming, he would just change producers. He would do this four or five times. He used to invent along the way, and would say, "We need a new idea," or wait for an idea. To me it was frightening working in this way. He would phone me during the night, and take me out to clubs where the bare-bottomed dancers were and talk to me about work, undoing all that I had accomplished during the day. My preparations happened in the street on location like on motorways or airports. I would just wait with a photographer for the people to arrive. I carried files full of character photographs; Fellini would look at them, put them to one side, and then he might go back to them.

I got on very well with director Mauro Bolognini because we shared the same ideas. He had this particular ability to make the best out of nothing. In those days money was always scarce, and he would agree to make films on a very low budget, but this enabled him to have the freedom that he wanted. I would go to him and I would say, "I haven't got anything for this set, only a chair." And he would ask, "Is it the right chair?" If so, it would be enough for him. Or if I had only 15 suits for women and 15 suits for men and very little clothing for the rest of the cast, he would say, "It will be enough." This was his mantra. And that's what made working with Mauro such fun and so interesting. He was a person with a very strong imagination.

I had a beautiful relationship with Liliana Cavani. Our personalities are different but the aims were the same. For **The Night Porter**, we spent two months together, talking, looking at photographs, dressing actors, re-thinking designs. At that time she was very interested in image. My starting point was a great German photographer, with whose work I was very impressed, and then of course we looked at all the paintings of the pre-Nazi and Nazi period. We also kept in mind the work of George Grosz. It's always important to have a reference point from which to create a style.

From script to costume is a very slow process. When I read a story, I always start by envisioning the face of a character. In Italian cinema the face used to be extremely important—sadly, this element of filmmaking has been lost. To me, the study of a character may include a certain type of sleeve, a particular neckline, a tie, and a hairdo, but the face is fundamental. I try to get to know the actor or the actress, and then I begin to work from that image. I try to change a hairstyle, or the makeup, or the physical structure to bring the actor and the character together.

It is very difficult to start this process over, as I had to do for **The Damned**. I had prepared for seven months, thinking that Vanessa Regole would be the main character. Then, as we started to shoot the film, Ingrid Thulin took her place. Vanessa was ideal for the character of the 1930s, she had fine bone structure and thin lips. But Thulin was much heavier in the mouth, and had the physical appearance of someone from the 1950s. When I did the film test she wasn't happy, because I waxed her eyebrows and underlined her lips and gave her a different hairstyle. But I worked cautiously with her, like a psychologist. Day by day, from five o'clock in the morning,

1

2

3

4

(1–7) **Death in Venice**: Stills and sketch of some of the exquisite female costumes in Visconti's film (1–2, 5). Dirk Bogarde as the avant-garde composer Gustav Von Aschenbach (3), who escapes to a Venetian beach for some rest, but develops feelings for an adolescent Polish boy, Tadzio (4, 7). Tosi making adjustments (6). The designer finds it difficult to watch the films he has worked on, once they are finished: "There is one moment, in the morning at breakfast, when the family comes into the breakfast room and they sit down at a table. Once they are sitting, from the distance I can say yes, they are beautiful, but if I see them again, well…"

5

6

7

I kept trying until she was ready for the set. She needed a look of the 1930s, and I needed to help bring that look to life.

Designing costumes is like being a sculptor. After considering the actor's face, you must then design for his or her physical shape. The prevailing body shape changes about every eight years. We don't really notice it, but the physical preferences of an era—the look that is popular at that moment—can make contemporary actors very wrong for the characters they are to portray. When I saw the Albanian refugees recently arriving in Italy, I actually saw the faces of the Italians of the 1940s. They reminded me of the people from that era. It is very difficult today to find an actress who could portray an image of the 1950s. Today they are made of bones, while before, a woman had bigger breasts and more flesh. In fact, the sensuality of that time is gone. For one film with Fellini, he and I were in search of a woman like Sophia Loren or Brigitte Bardot. We looked everywhere, we looked in France, and all over Italy. But in the end we had to resort to padding, and although we did do the padding, even the arms were different, they were harder, so as you can see, the body does change.

Of course, if instead of Claudia Cardinale for **The Leopard**, I had had a real Sicilian woman, a brunette, dirty and with greasy hair, it would have been an advantage for the role. But I had Claudia Cardinale, who was certainly a beautiful woman from the south, but well-groomed, and this made a difference. I did put oil on her hair in the evening so that her hair would be greasy, and we tried to make her face vulgar, but she is an elegant woman. Visconti did not want Burt Lancaster as the lead. He wanted Laurence Olivier, who would have been terrible, when I think about it today. Because today you see Laurence Olivier as an actor of his time, but not Lancaster.

Lancaster can go beyond any fashion, he is a physical, physiological, and natural force. At the beginning Visconti studied him, they studied each other, and Lancaster kindly asked Visconti for information on how he could be an Italian nobleman. Visconti used to say, "Don't worry about it. I will see to it when we work together. You must not worry." Lancaster wasn't happy about this, because he expected more direction. But then one day when I was doing his makeup, Lancaster said, "It's really stupid of me to ask Visconti what a real Italian nobleman looks like, because I have one (Visconti) before my very eyes."

If you see **Anna Karenina** with Garbo, designed by Adrian, it is a very beautiful film, but the costumes follow the fashion of the time. It is just this that differentiates it from **The Leopard**, which is not a fashionable film. This is due to the basic distinction between a fashion designer and a costume designer. A costume designer's work is from story and culture, while the fashion designer captures trends that are in the air.

There are some actors, stars, who are great personalities, and they always remain themselves. And their good fortune is to project that special, unique image. My favorite actors, however, are those who have chameleon-like qualities, who can transform themselves not only psychologically but also physically. One example is Robert De Niro, especially when he was young. In Italy, such an actor was Gian Maria Volonté, who could do an incredible job becoming a character, before even meeting with the costume designer. Of course this is a great help. Sophia Loren too was able to become a character, although it was not easy, because she really was a great personality as well.

The director Liliana Cavani, known for films and television documentaries that tackle socio-political issues, began a working relationship with Tosi in the mid-1970s with **The Night Porter** (2–4). Tosi worked closely with Cavani on developing a visual style for the then highly controversial film, referencing Nazi period paintings and the work of George Grosz. Painted preparatory work for **Beyond Good and Evil** (1), Cavani's biopic of the philosopher Friedrich Nietzsche.

1

2

3

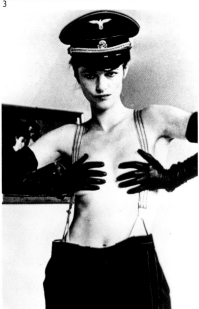

4

Stills from **The True Story of Camille**, a tragic love story (1–2) and **The Lovemakers** (3). Both films were directed by Mauro Bolognini, another prolific Italian director whom Tosi had a strong relationship with. "I got on very well with Mauro Bolognini because, apart from our friendship, we had the same ideas. He had this particular ability to make the best out of nothing." **The Lovemakers** was made in 1962; Tosi's costume and production design succeed in creating an astonishingly authentic picture of the latter half of the 19th century.

I had quite a time working on **La Cage aux Folles**. I expected a lot from Michel Serrault, whom I had not seen in the television show, so I didn't know how he would interpret his character on screen. I made him try costumes on in the fitting room until he became more and more awkward, and he looked at himself sad and mortified. He said, "Well, nothing is working." I was in despair. I was dressing him in pink, light green, but nothing happened. We used wigs, makeup, but nothing worked. He was just a big, sad guy. And then a miracle happened on the set. Before filming, he cried, "Ahhh!"—a great scream. He got rid of all his complexes and started to fly... But comedians are often like this. I did not have much fun making this film. In comedies, you must stress the character by helping the actor employ his uninhibited nature through his costumes.

A good relationship with the cinematographer is essential, as he can make a great contribution both to the director and the costume designer. There are costumes that have been rescued by photography. The cinematographer and costume designer are the same team: they must work together to set forth the color palette of a movie, expressing shapes with color. Most of the time I worked with the same cinematographers, Peppino Rotunno and Armando Nannuzzi. I would make suggestions about what I would like them to stress. I might have said, for example, "You can switch everything off, dim the light, but this particular tone serves as an important symbol, so please make it stand out." It would be an adventure, because the cinematographer would have to juggle the light with everything he was filming.

I don't know if there is a type of project that I most enjoy. Designing for films takes a special awareness. You must

always remember that your work will be seen in many different shots, that it will be broken into pieces. This is very different from the theater, where the audience gets a complete, full image of the costumes when the curtain goes up. If I manage to express myself within the design, it somehow belongs to me, and it is more pleasant to work. I also obtain better results. Somehow if I can participate in the design process, then I'll find pleasure. After months of research and work I do enjoy myself, but I admit I have to overcome my initial reluctance to make the commitment. Oddly, that discomfort reoccurs when the film is over. I never watch my movies because I want to change everything when I see them.

Perhaps the essence of costume designing is the willingness and humility to accept each project as a new venture, to bring no preconceptions to the work, and to accept that each film is a learning process. Costume design is about working closely with the director. In film the costume designer does his job, but ultimately, it is the director who must make the choices. He must use the costumes and the other production elements to illuminate the fictional world being created, and to put it into a dramatic perspective.

3

biography

Theadora Van Runkle is a natural. With no formal training in costume design, but working as an illustrator of costume, she got the job of designing **Bonnie and Clyde** (Arthur Penn, 1967) on the enthusiastic recommendation of designer Dorothy Jeakins. For her stunning, and then trendsetting costumes, she received an Academy Award nomination for this, her very first film. After this movie, Van Runkle went on

theadora van runkle

to Norman Jewison's **The Thomas Crown Affair** (1968), and she designed costumes for Elia Kazan's **The Arrangement** (1969), Peter Bogdanovich's **Nickelodeon** (1976), Martin Scorsese's **New York, New York** (1977), and Blake Edwards' **S.O.B.** (1981). Van Runkle was subsequently nominated for two other Academy Awards, both in connection with Francis Ford Coppola on **The Godfather, Part II** (1974), which is acknowledged to be one of the finest and most beautiful American films, and the good-looking, and lighter, **Peggy Sue Got Married** (1986). Her spectacular gowns have been worn by Lucille Ball in **Mame** (Gene Saks, 1974) and Dolly Parton in **The Best Little Whorehouse in Texas** (Colin Higgins, 1982). Van Runkle was honored with the Costume Designers Guild Career Achievement Award in 2002. Van Runkle's great gift has always been her ability to communicate through her detailed and sensitive renderings. Her eye for color and silhouette has been demonstrated by her fabulous costumes, while her drawings and paintings are highly prized internationally.

interview

My career as a costume designer began as a fluke. I was about 27 years old, and although I wasn't exactly a hippie, I certainly wasn't in the mainstream of the business world. The only clothes I had ever made were some things for my daughter and myself, and I had no experience in design or theatrics. I had been a commercial illustrator and times were tough, because advertisers had started using photographs rather than artwork. I felt my career was going downhill very rapidly. Then Dorothy Jeakins called me. She told me about a film she was about to start working on and said, "I've also been asked to design a little cowboy movie at Warner Brothers, and I have recommended you to do it. I know with your great valor that you can." I said, "But I don't know how to design a movie," and she said, "You can do it." I interviewed with Warren Beatty and Arthur Penn and got the job designing the costumes for **Bonnie and Clyde**. It was a time of great change in my life. The minute I opened the script, I saw everything that I would create, and I never really deviated. I knew it would be a great hit and that I would be nominated for an Oscar. So my start in film really derived from my talent as an illustrator, not as a dressmaker. I had no education or schooling in creating clothing. Somehow, I knew intuitively what shapes pieces of fabric should be cut in, and how they would hang and drape. It was something that I'll never be able to explain or understand.

I was influenced, probably like most costume designers, by the designers of the Golden Age. I loved Adrian when I was a little kid. I had seen **The Women**, and it was unforgettable the way Norma Shearer pinned that beautiful white shell on her gray flannel lapel. It changed my life. There was a silent movie house in my town, and when I was in my late teens I saw a Rudolph Valentino movie and a Dietrich movie there. The Dietrich film was **Shanghai Express**, and I was so overwhelmed by her beauty and by the costumes that Travis Banton designed for her. Banton designed wonderful things like long black-and-white reversible beach capes and espadrilles. They were so dramatic. I also loved the work of Gwen Wakeling. The films of this era were all I knew. I hadn't seen many of the movies of the 1950s and '60s when I got started in the movie business.

Someone once said to me that before **Bonnie and Clyde** came out there hadn't been any influential film costume work done for about a decade. I've always been intrigued by the idea that I had brought a painterly or illustrative approach to costume design that stemmed from an earlier era. Perhaps there is a unique quality to my work that is apparent to others but not to me. My daughter was once watching a movie on television that I designed, and I don't know how, but she knew that I had worked on it. I don't think that my designs have a definitive element or style, and I don't know what I would call my "trademark." To me, that is an indefinable quality.

My design process invariably began with sketches. The relationship between sketches and the final costume was very important to me, and sketching was the only way I could work. The idea came to me, and I had a drive to manifest it on paper. I often had sort of a Russian poster image in my mind when I was working. For example, I would envision a black-and-white costume against a sea of red in the background, or if it were a red costume, it would be outlined in black and white.

I would always read the script and make a tiny drawing of what the actor would be wearing next to the dialog. I would see the costumes in my mind's eye and then incorporate all the variables that go into making a costume actually work, such as the actor's physique, and the mood, location, and lighting of the scene. Would the director like it? Would the actor understand it? Would the leading lady look so alluring that the leading man would protest that she was stealing the scene? Somehow the costume had to address all these issues, and in the course of making a final presentation drawing, I had to keep all of them in mind.

When designing, I always had to consider the issue of finding the right fabric. To realize the vision, the fabric couldn't be too soft, and it couldn't be too stiff. If it was supposed to be stiff it had to have substance, yet it couldn't be so thin that it wouldn't flatter the actor. I tried to work all these factors out in the course of the drawing. I didn't generally make too many mistakes in the actual cutting and fitting, so the thought that went into the design was worth it in the end. This was my method early in my career, when all of the costumes for a movie—both contemporary and period films—were made, rather than bought off the rack at a department store. I would put together a huge portfolio to show an actor, including

1

2

3

4

5

6

(1–7) **New York, New York**: There was so little money available for Martin Scorsese's 1977 critique of the MGM musical that production designer Boris Leven had to make one of the trains in the film out of cardboard boxes. In stark contrast to the miniscule budget, many of the film's 250 dress extras wore gowns by Adrian, loaned by private collectors. "I don't know if it even showed, but it was a good spine for the power of the whole movie just to have it there." Stills and corresponding sketches of Liza Minnelli and Robert de Niro's costumes in the film (1–6). Scorsese (right) with Van Runkle and Minnelli (left) on set (7).

7

1

2

3

(1–9) **Bonnie and Clyde**: The 1967 film, directed by Arthur Penn, for which Van Runkle was Oscar-nominated, is a stunning recreation of Depression era 1930s America. This was Van Runkle's first film—she had previously been working as an illustrator—and she cites it as one of her favorites, due to the amount of creative freedom she was given. She also felt very lucky to have actress Faye Dunaway as her first leading lady, a "ravishingly beautiful" woman with a naturally sensual body shape (2–6). Sketches for Dunaway and Warren Beatty's characters (1, 8–9), and a photo of Dunaway and Van Runkle in Bonnie's trademark beret, promoting the film (7).

4

5

6

8

Bonnie and Clyde Fashion Show

Theadora Van Runkle will be guest moderator at the Century Square Bonnie and Clyde fashion show at 2 p.m. Saturday. Miss Van Runkle (to right of Fay "Bonnie" Dunaway) has been nominated for an Academy Award for costume design prepared for Bonnie and Clyde. Fashions of the 1930s will be displayed during the half-hour showing.

7

9

(1–5) **The Godfather, Part II**: Director Francis Ford Coppola had wanted Van Runkle to design for the first film in the trilogy, but she had declined after finding some parts of the script too violent. He gave her complete freedom on the sequel. She manufactured all the costumes in the film, and found that the cast, including Al Pacino (3), had an immense amount of trust in her. The party scene: Coppola had wanted the men to be wearing tuxedos, but Van Runkle had put them in gray silk sharkskin suits. After an angry 15 minutes, the director changed his mind (2). Two of Van Runkle's drawings from the film (4–5). Despite the length of the film, there were many scenes that hit the cutting room floor, which included female characters Van Runkle had designed, one of whom had a staggering 57 costume changes.

perhaps 150 finished, painted drawings. In retrospect, this was not the most efficient method. If I'd only known at the time how Adrian managed this. He would make many little drawings for each possible costume and he would show those to the actor. The actor would pick out the ones (maybe two or three) that he liked best, and Adrian would do his final drawing based on the selections.

One problem with showing sketches for an actor's approval is that many actors don't have a talent for actually *seeing* a drawing. For **Mame**, for example, I remember showing Lucille Ball a huge portfolio of colorful costumes in violets and fuchsias, black and white, brilliant reds, and the tones of brown and beige that I thought would become her. When she finished looking at the portfolio she said, "Does everything have to be black and white?" She didn't know she'd been looking at a lot of color. I worked with Lucy's wardrobe women on this movie, and they were wonderful. I could say to one of them: "The sleeve. I need more soul in the sleeve, and I'm going to come back after lunch and I want to see soul." And I would come back and they would have done it. They were so wise and so devoted. It was just heavenly. Lucy had opinions that she didn't hesitate to share with me, but in the end she would follow my advice. She was very patient in her fittings, and she would stand for as long as I needed her to. I was so single-minded and driven—I didn't make it a party. It was work, but she must have trusted me in some way. It was because of Lucy's looks that the clothes worked. It was the same with Faye Dunaway. How could I have been so fortunate to have someone as ravishingly beautiful as Faye Dunaway for my first leading lady? She was the perfect choice for that character. You really felt that she was naked under the clothes. I think it was the first time since Jean Harlow that somebody on screen was nude beneath her costume.

When I started as a costume designer, designers generally had much more time to prepare for a picture and a bigger budget than they do now. On the other hand, they usually had limited access to the actors. Cast members would often be shooting on location and weren't available for fittings during pre-production. Early in my career, getting work was much less formal than it is now. A designer would meet with the director, and if he took a shine to you he would hand you the script, and you had the job. Now the process is very long and attenuated. Before I worked on **The Thomas Crown Affair**, I joked around with director Norman Jewison for about five minutes and he said, "Well, are you any good?" and I said, "Yeah," and he said, "Okay. Here's the script." I went home and read it and started to work. I had a lot of self-confidence, and I think that it worked in my favor. For **The Thomas Crown Affair**, I simply created my sketches as I thought appropriate. I didn't discuss them with Norman, nor did I follow any hints in the script about what the heroine or hero was wearing.

I loved meeting Francis Ford Coppola and working on **The Godfather, Part II**. Francis was always very kind to me and he gave me a lot of freedom. I just read the script and put together my designs. I didn't consult with him about any of the costumes. On our first day on location, we were at Lake Tahoe shooting the party scene. I had dressed Al Pacino in gray club silk (we manufactured all the movie's costumes) and Francis expected everyone on the set to be in tuxedos. When he asked me why they weren't, I said, "It's an afternoon party and they're trying to become WASPs. They would never wear tuxedos." Francis turned around and walked into his cottage and slammed the door. He emerged about 15 minutes later and he said, "I thought it over and you're right." He was able to see

1

costume design

Script 27

3

2

4

(1–8) **Mame**: Morton Da Costa's Oscar-nominated 1958 film **Auntie Mame**, starring Rosalind Russell, spawned a Broadway musical and this lavish 1974 film adaptation. It starred Lucille Ball, the much-loved soap and film star, who was the first American woman to own her own film studio, called Desilu. Ball wanted all the costumes that Van Runkle had designed for her to be made in her own work room at Desilu. "She told somebody that she thought the clothes were the best she'd ever worn."

6

7

8

(1–6) **Peggy Sue Got Married**: Fourteen years after director Francis Ford Coppola first contacted Van Runkle to design **The Godfather**, they collaborated on this 1986 flashback film. Van Runkle was nominated for an Academy Award for the perfectly realized costumes from the 1960s and '80s (2–6). Van Runkle making costumes in the MGM workshop (1).

1

2

3

4

5

6

the characters deciding to dress in this way; trying to wend their way into the good graces of the community. I hadn't clearly thought of that, I just saw it intuitively. Being an artist himself, Francis was able to respect my decisions and my contributions. I can't imagine a director today giving a costume designer that much freedom.

Probably the most challenging film I designed costumes for was **New York, New York**. Our budget was so tight that when Liza Minnelli was in a scene at a train station, production designer Boris Leven had to make the train out of cardboard boxes. We had no money at all for costumes. I cut up some of Liza's Halston gowns, and I bought fabrics downtown for a dollar a yard. We strung our own beads. Toward the very end of filming the ending was rewritten and the plan was to shoot a very Hopper-esque scene, with Liza as an usherette. Irwin Winkler, the film's producer, called me in to ask me if I could get 15 usherette costumes made by the next day. I said yes, and he said, "I can only give you $15 a figure." With pantyhose at $6 a pair, I didn't know how I was going to do this. I didn't think I could find enough pillbox hats and little military-type jackets in stock. But I did it somehow.

It was a very demanding project with many costumed extras, and it seemed like we worked all night, all the time. Being a costume designer requires a great deal of energy, both physical and emotional, and working on the set is a lot of responsibility. My greatest fear was always that I would delay the filming. This happened only once, on **New York, New York**. We planned to film the big finale scene where Liza sings *New York, New York* on a Monday. I had designed a black, bugle-beaded gown for the scene. I kept fitting her in it to make sure it would work. As weird as this may sound, I had been tracking Liza's weight during the filming and I knew that she was not at her slimmest on Monday mornings. I worried about it all weekend—and rightfully so, because on Monday morning I couldn't do up Liza's zipper. The director, Marty Scorsese, said to me, "Well, you figure it out. We've got 250 dress extras on the set." It was 9:45am and I called for a limousine, which arrived immediately. I was praying that some of the shops would be open by 10:00am. I had $300 in cash on me. We pulled up in front of some shop and I leapt out. Inside I saw a great big red smock dress of Halston's. I literally threw the money across the counter and got back to the studio. Although the experience on **New York, New York** was quite harrowing, one thing I loved about being a costume designer was thinking on my feet. Having the opportunity to creatively solve problems was always exciting.

My passion for fabrics is what made designing costumes such a joy for me. I loved fooling around with material and seeing what it would do. I loved comparing colors and matching patterns, seeing what I could create with a plaid and a paisley. That kind of fiddling around was always my favorite aspect of the job.

I feel so fortunate that I was able to design costumes for a living. As a costume designer, I think that my ability to express my ideas through drawings was my greatest asset. When I was about six years old I used to lie in bed on weekend mornings and dream of painting immense murals. One day I was expressing to a friend my disappointment about having never created huge paintings, and she responded, "Are you kidding? Through the cinema you have made the biggest murals there are."

With over 60 film credits to his name, Albert Wolsky is one of the most prolific costume designers working today. Born in Paris, Wolsky later studied at the City College of New York and began his career in the theater. Wolsky's first movie was **The Heart is a Lonely Hunter** (Robert Ellis Miller, 1968). He has worked with countless directors, and is known for his collaborations with Bob Fosse and Paul Mazursky. For Fosse,

albert wolsky

163

interview

Wolsky designed the cool, early 1960s clothes for Dustin Hoffman's dark portrayal of comedian Lenny Bruce in the black-and-white **Lenny** (1974), and the Broadway chorus line, combined with the rich fantasy life of choreographer/director Fosse in **All That Jazz** (1979), earning Wolsky his first Academy Award. His diverse credits with Paul Mazursky include bittersweet, understated, contemporary comedies, and exquisite period melodramas. These include **Harry and Tonto** (1974), **Next Stop, Greenwich Village** (1976), **An Unmarried Woman** (1978), **Willie & Phil** (1980), **Moscow on the Hudson** (1984), **Enemies, A Love Story** (1989), and **Scenes from a Mall** (1990). Wolsky won another Academy Award for the costumes in Barry Levinson's **Bugsy** (1991), and was nominated for Levinson's **Toys** (1992). His ability to capture and project a film's tone is seen in all of his work, from his upbeat costumes for **Grease** (Randal Kleiser, 1978), to his understated costumes for **The Road to Perdition** (Sam Mendes, 2002).

I was always interested in design, but my pragmatic side had me working in my father's travel business after college. I was almost 30 years old when I had a crazy epiphany. I thought, "I love clothes and I love theater—why not become a costume designer?" A friend introduced me to the formidable Helene Pons, who had a costume house on 54th Street. She was about to start executing the costumes for the Broadway version of *Camelot*. I went to her to get some career advice, and to my amazement left with a job offer. It took me a while to realize she was very smart. She saw someone who managed an office of more than 20 people—so for $100 a week she got a manager. The arrangement was very beneficial for both of us, and the year I worked there was the beginning of my education.

After working for a few years as an assistant, I began slowly doing some off-Broadway and television work on my own. Theoni V. Aldredge, whom I had assisted on a Broadway musical, called about working on a feature film, **The Heart is a Lonely Hunter**. I was confused and thought she wanted

(1–6) **Bugsy**: Wolsky worked with director Barry Levinson and cinematographer Allen Daviau on this tale of 1940s gangster Benjamin "Bugsy" Siegel, played by Warren Beatty. Annette Bening starred as Virginia Hill (4–5). Wolsky's swatch board showing fabric samples of all Virginia Hill's outfits (3). Wolsky (center) dressing Bening on set (2). When he was designing her opening outfit (1), Wolsky worked right on the form, ignoring the grain and creating the shape using pins.

costume design

1

2

3

4

5

me to assist her, but she was desperately trying to arrange for me to interview for the costume designer's job. That was the beginning of my career in the movies. I've been helped along the way by many people, particularly other generous designers.

When I started designing costumes for Hollywood movies in the late 1960s and early '70s, a movie was driven by the vision of an individual director, such as Paul Mazursky, Hal Ashby, or Francis Ford Coppola. The climate of filmmaking has changed. Today, the studios have become a small part of the whole pie. So it all becomes about corporate decisions, which are made based on immediate accountant gratification. It is a climate that limits artistic vision. But, on the other hand, I don't see much change in my process of designing costumes. What costumes contribute to a movie, and how I go about designing them remains the same.

I'm very director-oriented. I don't need to be given exact information such as "the dress should be blue," but I need to know why a director likes a particular material. Going on location scouts is very valuable, as they give me an idea of the type of environment the director is looking for to set the story. I find that I rarely ask specific costuming questions. I listen carefully, and I get a sense of what draws a director and what repels him. To me, it's most difficult if I feel that I can't read a director. If I sense that I'm not in sync, I have to find a way to tune in to the director's style of communication.

I think that two of the most significant directors with whom I've worked are Paul Mazursky and Bob Fosse. Neither of them were "dictators" about design, but they worked very differently. Paul is relaxed and jocular, and he creates a comfortable environment. He'll know immediately if he likes a design, and he'll know why. The more we worked together, the easier it became for me to anticipate what he would want. That's the beauty of working with someone for a long time; you develop a form of shorthand.

My first film with Bob Fosse was **Lenny**. Designing costumes for this movie had a strong effect on me psychologically. We had no budget, and we were working 18 hours a day, six days a week. I barely had time to think before the next scene came up. When it was all over I felt like I had risen to a new level, both in terms of what I had learned and what I was ready to do. It was a turning point for me. Fosse was a mystery, and working with him was always intense. He would relate something specific in mind, but he delegated most decisions. For example, for **All That Jazz**, he decided that he wanted to see the veins on the dancers. I quickly had the veins painted onto the leotards, only to see the paint start bleeding once the dancer began to perspire. To solve the problem I decided to have all the veins appliquéd onto the fabric. That was a much better way of solving it, and it gave depth and texture to the effect. Learning from one's mistakes is part of moviemaking.

Designing costumes for a movie is a very collaborative activity; it's about being part of a whole process, with many contributors. I start every project by reading the script and reacting, emotionally and visually, to its characters. It's almost as if I have the rough sketches in my head and they simply pop out. I may start doing some research if the film is a period picture, but I need to talk to the director before I go any further. After the director, the three people who must work joined at the hip are the costume designer, the cinematographer and the production designer. It is up to all of

1

2

3

4

5

(1–8) **Grease**: Made in 1978 but set in 1959, this hugely popular movie musical perfectly captured the look of the late 1950s. The costumes for **Grease** relied heavily on a range of strong, primary colors. The classic dance-off (1–2) and the sexy Sandy, played by Olivia Newton-John (3). Wolsky's sketch for The Pink Ladies (8) and a watercolor sketch (7) for the Beauty School Dropout scene (6).

6

7

studies
Grease (1978)

PEDAL PUSHERS
BALLERINA
SLIPPERS

GREASED
CARDIGAN
T SHIRT

8

us to create the look of the picture. I'll always work closely with the cinematographer to get the right combination of colors. I like working in blocks of color, and I generally eliminate as much color as I can, because it gets heightened on screen. I tend to work with very muted tones that may look a little drab to the eye, but they won't look drab on-screen. They always warm up. It's the nature of film. To me, non-primary colors are the most effective, and I prefer to use subdued palettes, particularly rich browns and tobaccos and mauves. If something is red, I'd like it to be a brick- or a brownish-red. Of course, when a project calls for color, you have to let go, as I did for **Grease** for example.

The camera creates an ambience—it's almost a kind of smell to me. I need to have that sense of a movie's tone, which is why I hate pre-production. I just don't know what I'm doing until filming is underway. There's kind of an unreality to the pre-production stage. Once filming begins, I need to be on set at the start of everyday. I'm there to establish a scene, check on the background, which I am a bug about, and in general, get a feeling of the film's rhythms. Seeing rushes daily is also a necessity. All these elements give me my focus and reality.

A great amount of the film shot for a movie is never actually seen by the audience, and I never know what part of my work will show—will the viewer see only a character's back, bust, feet? Each part of a costume must work, and say something distinctive about that character, at a glance. This took a long time to learn, because my initial work was for the stage. I used to think "theater, far away; movies, close-up," but it's just the opposite. On film you don't get the image that your eye sees standing on the set, it flattens out. The stage affords a designer better control of what the audience actually sees.

You can see a material's texture better from the last balcony than in a close-up in a movie. A costume designer can never think something "won't show." A designer should love his work, including all the details of costumes, regardless of whether he thinks they'll be seen on the screen. I worry about every single item of clothing, down to the shoelaces. I wouldn't want anyone on my crew to learn any other way of designing.

For contemporary movies and period films, a costume designer needs a keen power of observation. The craft has a strong human element—what do a person's clothes tell you about them? Being observant. and being aware of the world around you is an important quality in a costume designer.

Designing contemporary films is difficult. My job is to identify, largely through elimination and simplification, who somebody is. A costume designer conveys the character, whether he's poor, he's had a good day, whether he's about to commit suicide. This goal becomes more and more of a challenge with contemporary clothes. Anything goes in fashion today, and when anything goes, a costume doesn't say anything. In contrast, with period costumes you can define someone's economic status relatively easily.

For period films, there are ways to identify and distinguish each period. For example, for **The Road to Perdition**, which is set in the early 1930s, I told director Sam Mendes that I was going to put everybody in hats. Of course in photographs of the time you will see some people bareheaded, but most people wore hats all the time. I deliberately put everyone in a hat. Emphasizing the traits of a period give a true feeling about it. Like all projects, **The Road to**

1

2

(1–5) **The Road to Perdition**: Wolsky went to great lengths to ensure the costumes were perfect. He often makes every part of a costume, but in this case went as far as designing the costume fabric itself for Jude Law (1), Tom Hanks, and the young Tyler Hoechlin (2). He relayed his instructions to the maker back and forth by telephone, as the schedule was too tight for a personal visit. The various samples and finished version of the plaid fabric for Michael Sullivan Jnr.'s Mackinhaw (4). Fabric swatches (5) for Paul Newman's character John Rooney (3).

4

5

3

(1–5) **All That Jazz**: This musical comedy was a multi-award winner in 1980, and won Wolsky an Oscar for Best Costume Design. It was the second time he had worked with director Bob Fosse, who had previously directed **Lenny**. Still and sketch of fan dancers (1–2) and stills from two of the film's musical numbers (3–4). Wolsky created this "vein" effect by appliquéing red and blue piping onto the dancers' leotards (5).

Perdition had its own set of challenges. I couldn't find the right fabric for the film's costumes. Nobody makes 1930s-weight, heavy woven wool anymore, but we finally found a hand-weaver in upstate New York who wove the fabric for us. I've made shirts, ties, suits, hats, even shoes, but this was the first time that I made fabric for a costume. In addition, everybody in this gangster movie seemed to get shot, even the minor characters. With all the blood and shotgun squibs, I needed multiple costumes, so couldn't rely on stock wardrobe.

One character's look can do a tremendous amount to define a period in a film. When I designed the costumes for **Bugsy**, I was eager to see one character in an entire shot, head to toe. One day the cinematographer, Allen Daviau, took me aside, and told me that he'd given me a present that day. It was a wonderful full figure image of Annette Bening walking toward Warren Beatty with the light behind her, coming through her long dress. It was a great way to set the tone of the scene and at the same time use costumes to convey a sense of the 1940s. When I'm researching period projects, I'm definitely influenced by painting and photography. I love this part of the job. I don't look at just how clothing was cut and the width of the shoulders. Part of a period's style is how people wore clothes. What was a woman's best dress? What was her worst dress? What were the norms of the times? I'm curious about people, what they wore, and why they wore it. When I know why someone chose something, then I know what to do as a designer to recreate a scene, a moment, a time.

One of my favorite films was **Enemies, A Love Story**, which was directed by Paul Mazursky. The characters were very familiar, and I felt that many of the costumes came from my personal memories. I understood the texture of the story's

time and place. My mother wore a suit just like the one Anjelica Huston wears in the film. I draw from many images from the past when I work. Early in my career I designed costumes for a movie called **Where's Poppa?** with Ruth Gordon. I recalled an image of a group of little Haitian girls going to a party in their fancy dresses, all with sweaters underneath because of the cold. I put a sweater underneath Ms Gordon's nightgown when she went to bed, and in its odd way it worked wonderfully.

On rare occasions, I have had a problem with an actor. I freeze when I'm working with someone who is inflexible or not collaborative. I won't ask an actor to be uncomfortable. In fittings, it's my job to solve a costume issue with an actor. I'm more likely to disagree with an actor over modern than over a period costume. Some actors don't have a strong concept about creating and building character through clothes. They just want to look pretty that day. If an actor believes that he "can't" wear something, he'll be right. It's not going to work. It's paramount that the actor walks on the set feeling comfortable and able to be visually in character. With a true superstar, I try to look for a way to be true to character, while retaining something about who they are as a personality. Katharine Hepburn always looked like Katharine Hepburn, no matter what she wore. But, I've also worked with many actors whose creative input have been a tremendous influence in the final outcome. Those are the joyous moments.

Although I was never trained to draw, I don't generally work with sketch artists because I have a hard time relating my images in words. **Galaxy Quest**, however, was set in space,

albert wolsky

costume design

(1–4) **Enemies: A Love Story**: Wolsky felt a resonance with the personalities and setting of this film in New York, 1949—"A lot of them could have been my relatives, I understood the texture." Particularly in the case of Anjelica Huston (1–2): "My mother wore a suit just like the one Anjelica wears in the film."

2

3

4

and the costumes had a certain fantasy element to them. I realized that precise, illustrative sketches were going to be important for the director and the producer, so I had a sketch artist help me. For years I thought that anybody who could draw was a brilliant designer. It took me a while to be able to look at a sketch and think: "This is a gorgeous drawing, but it's not a good costume." The costume has to support the character and the story, it has nothing to do with being attractive or day-to-day fashion. Costume design is about what people wear, why they're wearing it, what their class is, and what their needs are. It's to tell a story. Just as the actor is telling a story, the director is telling the story, a costume designer is helping to tell the story visually. His costumes should instantly tell the audience who someone is, using no dialog.

My objective as a designer is to work with full integrity, despite whatever obstacles may appear. I must guard against sliding into making do, or taking the easy way out. On my first movie, I watched cinematographer James Wong Howe working at the end of his career, being as painstaking and worried as if it was his first film. That was a lesson I never forgot. It's the only way to work.

1

2

(1–2) **Galaxy Quest**: In the work room and at wardrobe tests for the science-fiction comedy. Although an uncharacteristic genre for Wolsky, he enjoyed the challenge of working on the film.

picture credits

Courtesy of The Ronald Grant Archive: p12 **The Last Emperor**, Recorded Picture Company; p15 **Brazil**, Universal Pictures (1–4); p16–17 **Dangerous Liaisons**, Lorimar Film Entertainment/Warner Bros. (1–7); p18–19 **The Last Emperor**, Recorded Picture Company (1–5); p22–3 **Restoration**, Oxford Film Company/Miramax (1–4); p27 **A Clockwork Orange**, Warner Bros. (1, 3); p28 **The Cotton Club**, Paramount Pictures/Orion (1–4); p30 **Barry Lyndon**, Warner Bros. (1); p31 **Barry Lyndon**, Warner Bros. (4–6); p33 **Out of Africa**, Universal Pictures (1–2, 4); p34 **Dick Tracy**, Touchstone Pictures, photography by Peter Sorel (1–3); p38 **Malcolm X**, 40 Acres and a Mule Filmworks/Guild Film Distribution (1); p39 **Malcolm X**, 40 Acres and a Mule Filmworks/Guild Film Distribution (2, 5); p40 **Do the Right Thing**, 40 Acres and a Mule Filmworks/Universal Pictures (1–4); p42 **Amistad**, Universal Pictures/Dreamworks SKG, photography by Andrew Cooper (3–5); p43 **Amistad**, Universal Pictures/Dreamworks SKG, photography by Andrew Cooper (6–8); p44–5 **Jungle Fever**, 40 Acres and a Mule Filmworks/Universal Pictures (1–4); p54 **Bram Stoker's Dracula**, Columbia Pictures (2); p58 **Erin Brockovich**, Columbia TriStar; p61 **Mighty Aphrodite**, Miramax (4); p62 **Bullets Over Broadway**, Miramax (1–2); p66 **Erin Brockovich**, Columbia TriStar (3); p67 **Erin Brockovich**, Columbia TriStar (6–7); p68 **Broadway Danny Rose**, Orion (1); p72 **The Blues Brothers**, Universal Pictures (1); p73 **The Blues Brothers**, Universal Pictures (5); p79 **Oscar**, Touchstone Pictures, photography by Andrew Cooper (2–3); p80 **Raiders of the Lost Ark**, LucasFilm/Paramount Pictures (1, 3–4); p82 **The Age of Innocence**, Cappa Productions/Columbia Pictures; p84 **The Adventures of Baron Munchausen**, Columbia Pictures (1); p85 **The Adventures of Baron Munchausen**, Columbia Pictures (2–3); p86 **The Age of Innocence**, Cappa Productions/Columbia Pictures (1–2, 5); p87 **The Age of Innocence**, Cappa Productions/Columbia Pictures (6–8); p88–9 **Indochine**, Orly Films (1–4); p90 **A Midsummer Night's Dream**, Fox Searchlight/20th Century Fox (2, 4); p92 **Tess**, Renn Productions; p95 **Buffalo Bill and the Indians**, De Laurentiis/EMI (3); p96 **Tess**, Renn Productions (3); p97 **Tess**, Renn Productions (5); p99 **Papillon**, Allied Artists (1, 4); p100 **Death on the Nile**, EMI (1, 3); p101 **Death on the Nile**, EMI (5, 7); p102 **101 Dalmations**, Walt Disney (1); p104 **Shakespeare in Love**, Miramax/Universal Pictures; p107 **Shakespeare in Love**, Miramax/Universal Pictures (1–4); p109 **Velvet Goldmine**, Miramax, photography by Peter Mountain (1); p109 **Velvet Goldmine**, Miramax (2–3); p110 **Velvet Goldmine**, Miramax (1–3); p112–13 **Orlando**, Adventure Pictures (1–6); p114–15 **Orlando**, Adventure Pictures (7); p116 **Batman**, Warner Bros./D C Comics; p119 **Dune**, De Laurentiis (1–2); p120 **Batman Returns**, Warner Bros. (1); p125 **Demolition Man**, Warner Bros., photography by Andrew Cooper (2–3); p126 **Excalibur**, Warner Bros. (1–2); p128 **Places in the Heart**, Columbia TriStar; p131 **Places in the Heart**, Columbia TriStar (1, 3); p132 **Silkwood**, 20th Century Fox (2); p132 **Klute**, Warner Bros. (3–4); p134 **Midnight Cowboy**, Florin Productions (1–4); p135 **Midnight Cowboy**, Florin Productions (5–7); p136 **Working Girl**, 20th Century Fox, photography by Andy Schwartz (1–3); p142 **The Damned**, Eichberg Film (2, 4); p144 **Death in Venice**, Alfa Film/PECF (3–4); p145 **Death in Venice**, Alfa Film/PECF (7); p147 **The Night Porter**, Lotar Film Productions (2–4); p150 **Bonnie and Clyde**, Warner Bros.; p153 **New York, New York**, Chartoff-Winkler Productions (2, 4, 6); p154 **Bonnie and Clyde**, Warner Bros. (2–4); p155 **Bonnie and Clyde**, Warner Bros. (5–6); p156 **The Godfather, Part II**, Paramount Pictures (1–3); p158 **Mame**, ABC/Warner Bros. (1); p159 **Mame**, ABC/Warner Bros. (5, 8); p160 **Peggy Sue Got Married**, Zoetrope (2, 4–6); p166 **Grease**, Paramount Pictures (1–3); p170 **All That Jazz**, 20th Century Fox/Columbia Pictures (1, 4); p170 **All That Jazz**, 20th Century Fox/Columbia Pictures, photography by Alan Pappe (3); p172 **Enemies, A Love Story**, Morgan Creek Productions (3).

Courtesy of AlbumOnline: p6 **The Leopard**, Titanus (also on p141); p122 **Artificial Intelligence: A.I.**, Dreamworks SKG (1–2).

Courtesy of PA Picselect: p20 **Spider-Man**, Columbia Pictures (1–3).

Courtesy of Katz Pictures: p105 portrait shot, photography by Polly Borland.

Visual material contributed by James Acheson: p13 portrait shot.

Visual material contributed by Milena Canonero: p24 **Barry Lyndon**, with thanks and acknowledgement to Warner Bros.; p25 portrait shot; p27 (2, 4), with thanks and acknowledgement to Warner Bros.; p30 (2–3), with thanks and acknowledgement to Warner Bros.; p33 (3), with thanks and acknowledgement to Universal Pictures; p35 (4), with thanks and acknowledgement to Touchstone Pictures.

Visual material contributed by Ruth Carter: p36 **Malcolm X**, with thanks and acknowledgement to 40 Acres and a Mule Filmworks/Guild Film Distribution; p37 portrait shot; p39 (3–4), with thanks and acknowledgement to 40 Acres and a Mule Filmworks/Guild Film Distribution; p42 (1–2), with thanks and acknowledgement to Universal Pictures/Dreamworks SKG; p43 (9–10), with thanks and acknowledgement to Universal Pictures/Dreamworks SKG.

Visual material contributed by Eiko Ishioka: p46 **The Cell**, with thanks and acknowledgement to New Line Cinema; p47 portrait shot, photography by Brigitte Lacombe; p49 (1–3), with thanks and acknowledgement to Warner Bros.; p50–1 (1–7), with thanks and acknowledgement to New Line Cinema; p53 (1–4), with thanks and acknowledgement to New Line Cinema; p54 (1, 3–4), with thanks and acknowledgement to Columbia Pictures; p55 (5), with thanks and acknowledgement to Columbia Pictures; p56–7 (1–7), with thanks and acknowledgement to Columbia Pictures.

Visual material contributed by Jeffrey Kurland: p59 portrait shot; p61 (1–3, 5), with thanks and acknowledgement to Miramax; p62 (3–6), with thanks and acknowledgement to Miramax; p63 (7–9), with thanks and acknowledgement to Miramax; p64 (1–5), with thanks and acknowledgement to Orion; p66 (1–2), with thanks and acknowledgement to Columbia TriStar; p67 (4–5), with thanks and acknowledgement to Columbia TriStar; p68 (2–3), with thanks and acknowledgement to Orion.

Visual material contributed by Deborah Nadoolman: p70 **Michael Jackson's Thriller**, with thanks and acknowledgement to Optimum Productions; p71 portrait shot; p73 (2, 4), with thanks and acknowledgement to Universal Pictures; p73 (3) photography by Timothy White, with thanks and acknowledgement to Universal Pictures; p74 (1–2), with thanks and acknowledgement to Universal Pictures; p74 (3), with thanks and acknowledgement to Warner Bros.; p76–7 (1–7), with thanks and acknowledgement to Touchstone Pictures; p79 (1, 4), with thanks and acknowledgement to LucasFilm/Paramount Pictures.

Visual material contributed by Gabriella Pescucci: p83 portrait shot; p85 (4), with thanks and acknowledgement to Columbia Pictures; p86 (3–4), with thanks and acknowledgement to Cappa Productions/Columbia Pictures; p87 (9), with thanks and acknowledgement to Cappa Productions/Columbia Pictures; p90 (1, 3), with thanks and acknowledgement to Fox Searchlight/20th Century Fox.

Visual material contributed by Anthony Powell: p93 portrait shot; p95 (1–2), with thanks and acknowledgement to De Laurentiis/EMI; p96 (1–2, 4), with thanks and acknowledgement to Renn Productions; p99 (2–3), with thanks and acknowledgement to Allied Artists; p100 (2, 4), with thanks and acknowledgement to EMI; p101 (6), with thanks and acknowledgement to EMI; p102 (2–4), with thanks and acknowledgement to Walt Disney.

Visual material contributed by Sandy Powell: p109 (4), with thanks and acknowledgement to Miramax.

Visual material contributed by Bob Ringwood: p117 portrait shot; p119 (3), with thanks and acknowledgement to De Laurentiis; p120 (2–4), with thanks and acknowledgement to Warner Bros./D C Comics; p122 (3–4), with thanks and acknowledgement to Dreamworks SKG; p123 (5–9), with thanks and acknowledgement to Dreamworks SKG; p124 (1), with thanks and acknowledgement to Warner Bros.; p125 (4), with thanks and acknowledgement to Warner Bros.

Visual material contributed by Ann Roth: p129 portrait shot, photography by Brigitte Lacombe; p131 (2), with thanks and acknowledgement to Columbia TriStar; p132 (1), with thanks and acknowledgement to 20th Century Fox; p135 (8), with thanks and acknowledgement to Florin Productions.

Visual material contributed by Piero Tosi: p7, with thanks and acknowledgement to Alfa Film/PECF (also on p145); p138 **Death in Venice**, Alfa Film/PECF; p139 portrait shot; p140 (1), with thanks and acknowledgement to Mega Film/Divina Film/Dieter Geissler Film Producktion; p141 (2–4), with thanks and acknowledgement to Mega Film/Divina Film/Dieter Geissler Film Producktion; p142 (1, 3), with thanks and acknowledgement to Eichberg Film; p144 (1–2), with thanks and acknowledgement to Alfa Film/PECF; p145 (5–6), with thanks and acknowledgement to Alfa Film/PECF; p147 (1), with thanks and acknowledgement to Lotar Film Productions; p148 (1–2), with thanks and acknowledgement to Opera Films; p149 (3), with thanks and acknowledgement to Titanus.

Visual material contributed by Theadora Van Runkle: p2, with thanks and acknowledgement to Warner Bros. (also on p154); p151 portrait shot; p153 (1, 3, 5, 7) photography by Bruce McBroom, with thanks and acknowledgement to Chartoff-Winkler Productions; p155 (7), with thanks and acknowledgement to Century City News; p155 (8–9) photography by Bruce McBroom, with thanks and acknowledgement to Warner Bros.; p156 (4–5), with thanks and acknowledgement to Paramount Pictures; p158 (2–4), with thanks and acknowledgement to ABC/Warner Bros.; p159 (6–7), with thanks and acknowledgement to ABC/Warner Bros.; p160 (1); p160 (3), with thanks and acknowledgement to Zoetrope.

Visual material contributed by Albert Wolsky: p162 **Bugsy**, with thanks and acknowledgement to Baltimore Pictures; p163 portrait shot, photography by Shonna Valeska; p164 (1–5), with thanks and acknowledgement to Baltimore Pictures; p166 (4) photography by Dave Friedman, with thanks and acknowledgement to Paramount Pictures; p166 (5), with thanks and acknowledgement to Paramount Pictures; p167 (6–8), with thanks and acknowledgement to Paramount Pictures; p169 (1–3) photography by Francois Duhamel, with thanks and acknowledgement to Dreamworks SKG; p169 (4–5), with thanks and acknowledgement to Dreamworks SKG; p170 (2), with thanks and acknowledgement to 20th Century Fox/Columbia Pictures; p171 (5), with thanks and acknowledgement to 20th Century Fox/Columbia Pictures; p172 (1–2) photography by Takashi Seida, with thanks and acknowledgement to Morgan Creek Productions; p172 (4) photography by Eric Leibowitz, with thanks and acknowledgement to Morgan Creek Productions; p173 (1–2), with thanks and acknowledgement to Dreamworks SKG.

index

costume design